NEW YORK STATE GRADE 5 ELEMENTARY-LEVEL MATH TEST

Rafael A. Mercado, M.S., ED., and Gabrielle A. Morquecho, M.S., ED.

BARRON'S

About the Authors

Rafael A. Mercado and Gabrielle A. Morquecho have over 30 years of classroom experience, mostly in the Buffalo Public School (BPS) district. Mr. Mercado has been a math teacher for the 7th–12th grades in the BPS and for the 5th–9th grades at the Gifted Math Program at the University of Buffalo. He is presently the associate director of the Gifted Math Program and a school administrator at Grover Cleveland High School in Buffalo, NY, and is currently working toward a Ph.D. in higher education. Ms. Morquecho has been a classroom teacher of special education students, is presently a school administrator at School #30 in Buffalo, NY, and is working toward her Ph.D. in learning in instruction.

Dedication

To Mom and Dad: thanks for your love and encouragement. To Luis and Alexandra: you are both funny, and I love you both very much. To Joshua, Tyler, and Jessica: each one of you has brought laughter in dark days, worries watching you grow up, and pride in the types of young adults you are turning out to be. I can't wait to see how your lives progress. To Gabrielle, my best friend and confidante: thanks for showing me kindness when I needed it the most. And, finally, to Patrick and Abuelita: I am sad that you didn't see the publication of my first book. I miss and love you both.

Rafael A. Mercado

To Lisa and Renee, my sisters and friends: thanks for all of your encouragement. To Diane: you have been there all of my life, sharing your wisdom and being a model for me. I love you for that. To Mom: thanks for your advice and love. And, to Dad: I truly, truly miss you.

Gabrielle A. Morquecho

All inquiries should be addressed to:
Barron's Educational Series, Inc.
250 Wireless Boulevard
Hauppauge, NY 11788
www.barronseduc.com

ISBN: 978-0-7641-3945-1
ISSN: 1940-4999

Date of Manufacture: January 2013
Manufactured by: B11R11, Robbinsville, NJ

Printed in the United States of America
9 8 7 6 5 4

10%
POST-CONSUMER WASTE
Paper contains a minimum of 10% post-consumer waste (PCW). Paper used in this book was derived from certified, sustainable forestlands.

Contents

IMPORTANT NOTE: Barron's has made every effort to ensure the content of this book is accurate as of press time, but New York State exams are constantly changing. Be sure to consult **www.p12.nysed.gov/apda/ei/eigen.html** for all the latest New York State testing information. Regardless of the changes that may be announced after press time, this book will still provide a very strong framework for fifth-grade students preparing for the exam.

HOW TO USE THIS BOOK

New York State schools begin the fall semester for the most part a few days after Labor Day. Typically, the Grade 5 mathematics exam is given in March. Given the fact that there are five areas of mathematical proficiency that the state of New York requires, there are about 38 days to prepare for each of these proficiencies.

To prepare for the exam, it is recommended that students spend about 10–15 minutes in addition to their regular homework time rotating among the topics they have learned in school. For students to be successful in the New York State assessments, it will take practice.

I have been a teacher of mathematics for about 20 years, as well as an administrator in an urban setting, and have found that the best ally students have is *time*. At the same time, time is their worst enemy. Practicing the types of questions they will face during the exam is just as important for students as their skill levels. By practicing the questions found in the book and studying the explanations, students will gain more knowledge about how to do problems.

This book will guide students through each item and standard they will need to know for the state exam. Each chapter was designed with questions after each skill is explored. The answers to these practice questions along with their explanations appear at the end of the book.

At the end of the book there are also two full-length exams with answers and explanations. After they have reviewed the entire book, students should use one of the exams to gauge their learning and the areas requiring more skill building. Students should correct their exams and make sure they understand why the answers are the right ones.

It is suggested that the second exam be taken about 1 or 2 weeks before the state exam under exam conditions—meaning that the test is completed in the amount of time required by the state.

TO PARENTS

Your child has a great learning coach in you! You do not have to be a great player to be a good coach. Imagine trying to coach Tiger Woods. His coaches have "big shoes" to fill, but remember that even Tiger Woods needs a coach. Therefore, do not worry. Your role is to provide encouragement and the necessary time "on task" (away from distractions) so that your child may work with the Mathematics strands developed by the state of New York.

Begin by making flash cards with your child. Work with definitions, shapes, formulas, the "dreaded" multiplication tables, and so on. Take out a calendar and plan to spend about 15–20 minutes each day discussing topics related to mathematics with him or her. This may be a challenge, but it can be done. Work together with your child to create a schedule and to follow this schedule in order to fully support your child and give him or her the best opportunity for success.

The way to help your child is by planning a daily, consistent, long-range review. You do not have to be afraid about not knowing mathematics, because just providing your child with the environment he or she needs to accomplish the task is enormously helpful. Here are some of the web sites you should visit with your child if you have more questions regarding the exam or about mathematical topics you would like to explore with them.

> New York State Education Department—sample test and past-year exams: *http://www.emsc.nysed.gov/ 3-8/math-sample/home.htm*

> National Council of Teachers of Mathematics—a wonderful free resource to help students with interactive web solutions: *http://illuminations.nctm. org/ActivitySearch.aspx*

Your son and/or daughter should take note of what topics prove to be most difficult when answering the review questions. When further assistance is needed, students can either go over the review books chapters and/or ask their teachers for assistance. This review book can be successfully used only when it is used in conjunction with other available resources. You are your child's first and best teacher. I would like to thank you for your help in this important process. Education has never been an entire country's responsibility as much as it is now.

TO STUDENTS

Every year when you walk into your classroom the teacher has to do certain things—make sure that all the students have supplies, that the room is set up to learn, and so on. Among the things they want to know is what you need to succeed in mathematics.

Many of the questions you will see on the state exam are similar to the questions you have seen during the past 6 years in school. Therefore, do not worry if you don't remember or are unsure of how to get the answer. Think about the question, go over the definitions, and let your teacher and your parents help you along the way. Your friend(s) can help if they are willing to stay focused to accomplish this goal. Begin studying the chapters in this book early in the year; have your parents and your teacher help you develop a schedule you can follow, but begin early.

If you work *diligently* (hard, consistently), you will perform better on these types of tests.

Before we begin, there are a few basic things you should know and understand before addressing the performance indicator (which is a fancy way of referring to the skills you are going to have). The basics will be reviewed first, and then you can apply them under exam conditions to provide a model of an assessment test question using these concepts.

Throughout the book you will recognize the practice questions by the box around the title. The answers to all the practice questions are at the end of the book.

Good luck!

THE NEW YORK STATE GRADE 5 MATH TEST

One of the most significant by-products of the No Child Left Behind (NCLB) legislation is that every New York State fifth-grade student, along with all of their third- to eighth-grade peers, has to be part of a midyear assessment (usually during the first or second week in March). Each child is expected to achieve at or above a level 3 score to be considered at grade level. (The levels range from 1 to 4.) Students need to understand the importance of this exam. The examples used throughout the chapters, as well as the questions in both full-length exams at the end of the book, follow the format of the questions on the state exams that students have taken in the past.

The exam is divided into two books. Book 1 consists of 26 multiple-choice questions. The questions in this section are scored either correctly or incorrectly, and each question is worth 1 point without partial credit given. Questions in this section have the following form.

You have a jar of pennies $\frac{3}{8}$ full, and then at the end of the month you add another $\frac{2}{8}$. How much of the penny jar is filled?

A. $\frac{5}{64}$

B. $\frac{5}{8}$

C. $\frac{1}{8}$

D. $\frac{6}{8}$

The answer is b, $\frac{5}{8}$. The question asks how much of the penny jar is filled. It was first filled $\frac{3}{8}$, and then an additional $\frac{2}{8}$. Therefore, $\frac{3}{8} + \frac{2}{8} = \frac{5}{8}$.

These types of questions require students to know the facts of the curriculum—the written course of study for the fifth-grade school year. These concepts and facts will be explained throughout this book, and students will be given many opportunities to sharpen their skills.

Along with this book, the state web site *http://www.nysed.gov* gives additional examples from both Book 1 and Book 2 and the types of questions students will encounter. These questions will give students additional practice in preparing for the exam.

In Book 2 of the exam there are two types of questions students are asked to solve: **Constructed-response** and **extended-response**. The constructed-response questions are worth *2 points each*, and the extended-response questions are worth *3 points each*. The exam booklet will inform the student which type of question they are answering by the number of points each is worth.

The following rubrics are used to score the questions in Book 2.

Questions with a 2-point scale would be scored on the following rubric:

2 Points	A 2-point response is complete and correct. This response <ul><li>shows that the student truly understands the skills needed for solving this problem.</li><li>shows that the student answered all the parts of the question and used appropriate operations to solve the problem.</li><li>tells/writes specifically how they got that answer.</li></ul>
1 Point	A 1-point response is only partially correct. This response <ul><li>shows that the student is beginning to understand the problem but has some minor errors in arithmetic as well as some misunderstanding of knowledge.</li><li>gives complete answers but has incorrect logical reasons as to why the problems was solved in a specific manner.</li><li>may contain a correct numerical answer but required work is not provided.</li></ul>

0 Points	A 0-point response • may contain an incorrect, irrelevant, incoherent response or contain a correct response arrived at using an obviously incorrect procedure. • although some parts may contain correct mathematical procedures, overall they are not enough to demonstrate even a limited understanding of the mathematical concepts embodied in the task.

Questions with a 3-point scale would be scored on this rubric:

3 Points	A 3-point response is complete and correct. This response • shows a complete understanding of the mathematical concepts and how to "attack the problem." • shows that the student has completed the task correctly, using mathematically sound procedures. • shows clear, complete explanations and/or adequate work when required.
2 Points	A 2-point response is partially correct. This response • shows only partial understanding of the mathematical concepts and how to "attack the problem." • addresses most aspects of the task, using mathematically sound procedures. • may contain an incorrect solution but provides complete procedures, reasoning, and/or explanations. • may reflect some misunderstanding of the mathematical concepts and/or procedures.

1 Point	A 1-point response is incomplete and exhibits many flaws but is not completely incorrect. This response • shows only a limited understanding of the mathematical concepts and/or procedures embodied in the task. • may address some elements of the task correctly but gives the incorrect solution and/or provides reasoning that is faulty or incomplete. • shows in the answer several flaws that indicate misunderstanding of the importance of the task, misuse of mathematical procedures, or faulty mathematical reasoning. • reflects a lack of essential understanding of the underlying mathematical concepts. • may contain a correct numerical answer but required work is not provided.
0 Point	A 0-point response • is incorrect, has no connection to the problem being asked and shows lack of understanding of the problem and how to go about starting and getting the answer. • is one in which some parts may contain correct mathematical procedures but overall shows a major lack of mathematical connection to the problem and how to solve it.

Rubrics give students the standards by which they will be asked to explain their responses and the particular score each response is given. It is good practice for students to compare their work with the rubric's explanation of a good answer. You will find that in Book 2 there are four constructed-response questions, as well as four extended-response items. For each of these questions, students are asked to elaborate on and explain their answers. The more sophisticated the answer, the higher the point value.

Students are also expected to show their work throughout this section of the examination. When students are asked to show their work, they are being asked to apply their skills and logical mathematical reasoning in attacking a particular problem and literally show how they got the answer.

The following question is a constructed-response question. Please read it carefully and consider the answers given.

PRACTICE QUESTION 1

Sponge Bob Square Pants has 150 clams he found on the bottom of the ocean. He gave all of them away equally to each of his 5 friends. How many did each of his friends receive?

A. 30

B. 50

C. 145

D. 155

(Answer on p. 169.)

The other type of Book 2 question is the extended-response question. Here is an example of that type of question.

PRACTICE QUESTION 2

In the Tour de France, Lance Armstrong (the only man to ever win the tour seven times in a row) rode the following distances each day.

Day	Distance (in miles)
1	$15\frac{3}{10}$
2	$14\frac{1}{10}$
3	$16\frac{3}{10}$

Part A: What is Mr. Armstrong's total amount of miles for the first 3 days of the Tour de France? Show your work.

Part B: In the same race, Mr. George Hincapie, Lance's teammate rode $13\frac{4}{10}$ fewer miles each day.

What is the total distance, in miles, that Mr. Hincapie rode during the same 3 days?

Part A:

_____

_____

_____

_____

_____

_____

_____

_____

Part B:

_____

_____

_____

_____

_____

_____

_____

_____

(Answers on p. 170.)

When New York State assesses students in March, the expectation is that the exam measures everything a fifth-grader has learned up to and including the March test date. Therefore, some of the content students are responsible for includes information from the previous year. This is why it is so important to review some of the items they have not touched on or reviewed since the fourth grade.

The following will give you an idea of the state's expectations for all fifth-grade students on one of the New York State Mathematics Learning Standards: Number Sense and Operation.

5.N.1	Read and write whole numbers to millions	Pre-test
5.N.2	Compare and order numbers to millions	

5.N.3	Understand the place value structure of the base ten number system:
	    10 ones = 1 ten
	    10 tens = 1 hundred
	    10 hundreds = 1 thousand
	    10 thousands = 1 ten thousand
	    10 ten thousands = 1 hundred thousand
	    10 hundred thousands = 1 million	

5.N.4	Create equivalent fractions (given a fraction)	
5.N.5	Compare and order fractions including those with unlike denominators (with and without the use of a number line). Note: Commonly used fractions such as those that might be indicated on a ruler, measuring cup, etc.	
5.N.6	Understand the concept of ratio	
5.N.7	Express ratios in different forms	
5.N.8	Read, write, and order decimals to thousandths	
5.N.9	Compare fractions using <, >, or =	
5.N.10	Compare decimals using <, >, or =	
5.N.11	Understand that percent means part of 100, and write percents as fractions and decimals	

This should give you an example of the skills and concepts students need to understand before the test in March. The other four standards are Algebra, Geometry, Measurement, and Probability and Statistics. In addition to the *Content Strands,* there are five *Process Strands*: Problem Solving, Reasoning and Proof, Communication, Connections, and Representation. Both sets of strands are meant to give each student the mathematical proficiency to function in the global market.

Chapter 2

GEOMETRY

BASIC GEOMETRIC TERMS

The simplest geometric term is a point. A **point** is a "spot" or location in space and is usually labeled with a capital letter, say T. (Remember it can be *any* letter.)

Example: • T

If you take two points and connect those points, the path connecting the points is called a **line**. A line extends to "infinity and beyond"—remember Buzz Light-Year. It is because of the property of infinity that you *never* know how long a line is. You will never measure a line. A line is represented in the following way.

Example:

Each line has a particular name, like you and me; the name of this line is $\overleftrightarrow{RM}$. A line can be represented (named) by two capital letters as we have done here.

Example:

$$\overleftrightarrow{RM} \quad \text{or} \quad \overleftrightarrow{MR}$$

The type of orientation of a line determines its characteristics. For example, a football goal post is an example of **vertical lines**. Likewise, when you are standing, your spine forms a vertical line.

A horizontal line goes from right to left, or from left to right.

At times, it is important to understand what happens to lines when they cross each other and what their **properties** are (things that you can say always happen). For example, here you have two lines, $\overleftrightarrow{XY}$ and $\overleftrightarrow{AB}$, that **intersect** (cross and meet at a point) at D. (Later we will discuss the types of angles that are formed between two intersecting lines.)

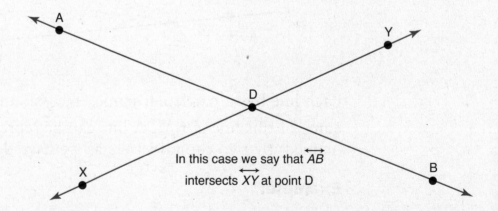

In this case we say that $\overleftrightarrow{AB}$ intersects $\overleftrightarrow{XY}$ at point D

Perpendicular lines—lines that intersect (meet) at a special angle to one another—that is, always a **right angle** (remember that right angles are made by lines that meet at 90°).

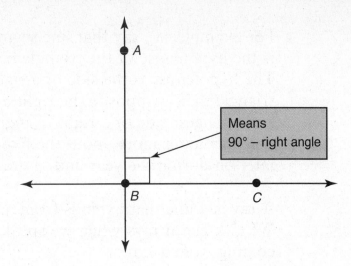

Means
90° – right angle

Parallel lines—lines that never intersect—keep the same distance between them, like railroad tracks. Can you find another set of parallel lines in this figure? We write this relationship as $\overleftrightarrow{AB}||\overleftrightarrow{CD}$. We say that line AB is parallel to line CD because $||$ means "parallel." Many times we need to use a line but use only part of it. The word "segment" means "part of." A line segment is shown as

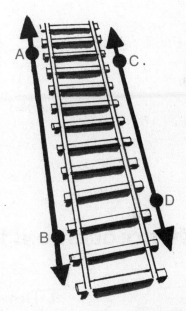

$\overline{AB}$, compared to $\overleftrightarrow{AB}$ for a line and $\overrightarrow{AB}$ for a ray. The only consideration you need to keep in mind for the ray is that the arrow above the letter B means that the ray started at a vertex—point A.

If we need to talk about geometric figures, we use the term **line segment** to denote the sides of these figures.

For example, we say that line segment $\overline{EF}$ is the hypotenuse of the triangle below. The **hypotenuse** is the side of a right triangle that is opposite the right angle. It is the longest side of a right triangle. You will encounter more about the hypotenuse later on during the year and in grade 6.

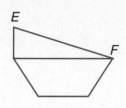

A **ray** is a line that extends from an endpoint to infinity. We talk about rays when we speak about rays of light coming from a star.

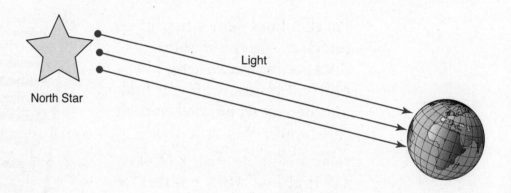

PRACTICE QUESTIONS 1

1. Draw a ray, a line, and a line segment.

2. Which types of lines do you see in this figure?

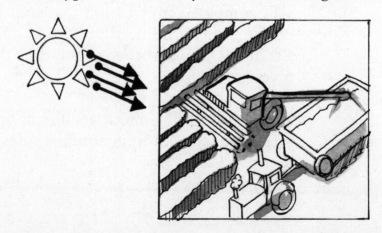

3. Count how many line segments you see in the following figure. Can you name three?

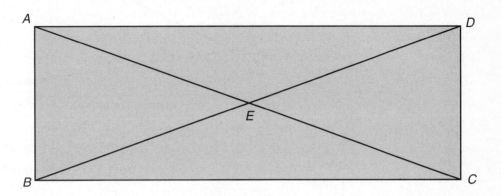

(Answers on pp. 171–172.)

ANGLES

This brings us to a point where we need to think about what happens when lines and/or line segments intersect each other. One of the things that happens is that they form **angles**. Angles therefore are said to form when two rays share a common endpoint. Point *A* in this case not only is the common endpoint of the two rays but also has a special name: **vertex**. Each angle has its own particular name. When naming an angle, however, you start from the letter at either end of the arrows, then name the vertex, and then name the other letter at the other arrow. So the written name of this angle is ∠*RAM*—which is read, "angle *RAM*." Since you can start at either of the two arrows, you could have also named this angle ∠*MAR*. A third way of naming this angle is ∠*A* because *A* is the vertex of the angle. When using the vertex to name an angle, the vertex is always placed in the middle.

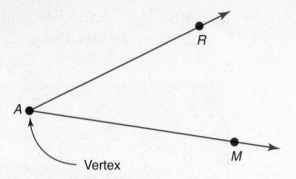

Angles are classified into one of four categories: less than 90°, equal to 90°, greater than 90° but less than 180°, or 180°. We give these angles specific names:

An angle that is *less* than 90° is called an **acute angle**.
An angle that *equals* 90° is called a **right angle**.
An angle that is *more* than 90° but *less* than 180° is called an **obtuse angle**.
An angle that *equals* 180° is called a **straight angle**.

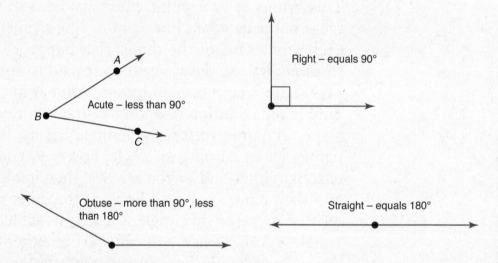

The New York State exam may ask you to identify these angles in a different orientation, so you may need to rotate your paper to figure out the type of angle the exam is referring to. Be ready to identify these angles regardless of their positions. The exam will also ask you to use a protractor to measure angles. How to use a protractor will be dealt with later in this chapter.

PRACTICE QUESTIONS 2

Classify each of these angles as right, obtuse, straight, or acute.

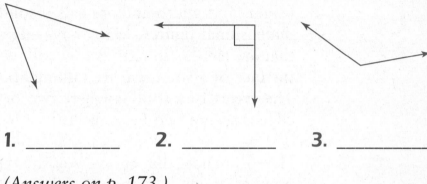

1. _____   2. _____   3. _____

(Answers on p. 173.)

TWO- AND THREE-DIMENSIONAL FIGURES

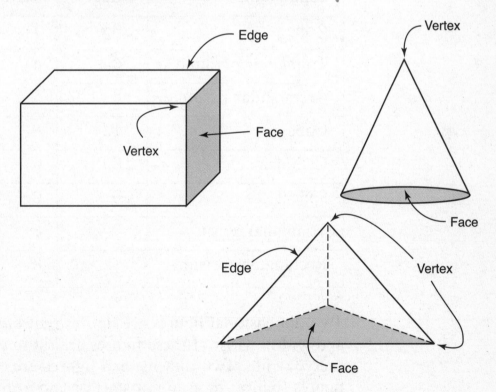

We are now almost getting to the "meat and potatoes" of the Grade 5 state curriculum, but we must first learn about three-dimensional (3-D) figures, sometimes referred to as space figures. You may think that 3-D is something foreign or strange, but it is not. Many of the objects you see every day are 3-D, such as cars, schools, your mother, your brother, your dad, your teachers, baseballs, and iPods. Things that are 3-D are objects that have length, height, and width and are called **solids** or **three-dimensional figures**. Many have edges, vertices, and faces that are similar to figures that follow. When we talk about the face of a solid, we are talking about the flat surface. The **vertex** of a solid is where two or more lines meet. An **edge** is where two faces meet.

Here is a chart that shows you the types of solids and the number of edges, vertexes, and faces each has.

Solid Name	Edges	Vertices	Faces
Cone	0	1	1
Triangular pyramid	6	4	4
Rectangular pyramid	8	5	5
Cube	12	8	6
Cylinder	0	0	2
Sphere	0	0	0
Triangular prism	9	6	5
Rectangular prism	12	8	6

Two-dimensional figures are figures that have only length and width. Most of these figures are those you see in everyday life. Two-dimensional figures are either **open figures** (figures that do not start or end at the same point) or **closed figures** (figures that have the same starting and ending points). We classify closed figures as **polygons** when they are made of straight lines. The word "polygon" comes from a Greek word meaning "many angles" (*poly*, "many"; *gon*, "angles"). With the exception of the triangle and the quadrilateral, every other type of polygon has a prefix referring to the number of angles, for example, pentagon, five angles, and hexagon, six angles. We usually use this information in addition to the number of line segments (sides) making up each polygon.

Name of Polygon	Number of Sides	Number of Angles
Triangle	Three	Three
Quadrilateral	Four	Four
Pentagon	Five	Five
Hexagon	Six	Six
Heptagon	Seven	Seven

Name of Polygon	Characteristics
Pentagon	Five sides and five angles
Hexagon	Six sides and six angles
Octagon	Eight sides and eight angles

If you are interested in continuing to investigate this information, go to the web site *http://en.wikipedia.org/wiki/Polygon*.

Before we take a look at all polygons, let us review the knowledge you are able to recall at this point.

Polygon	Characteristics	Example
Triangle	Three angles and three sides. If you add up all the angles of ANY triangle, the sum will equal 180°. Three types of triangles: • *Equilateral*—a triangle with all of its sides of equal length. • *Right*—a triangle with one of its angles equal to 90°. • *Isosceles*—a triangle with two of its sides of equal length and two of its angles of equal degree. • *Scalene*—a triangle with all of its sides and angles of different length and different degree.	Equilateral Right Isosceles Scalene
Quadrilateral	Closed figure with four sides and four angles. A quadrilateral can be a parallelogram or a trapezoid. Three types of parallelograms: • *Rectangle*—a parallelogram with four right angles and with opposite sides equal in length to each other. • *Square*—a rectangle with all sides equal in length. • *Rhombus*—a parallelogram with all sides equal and opposite angles equal in degrees; can be a square. One type of *trapezoid*—a quadrilateral with only one pair of sides parallel to each other.	Rectangle Square Rhombus Trapezoid

A fact that is very useful when thinking about quadrilaterals is that all the angles of a quadrilateral add up to 360°. This fact is often referred to in exams.

Is a trapezoid a parallelogram? Why? Why not? The following Venn diagram will help you remember the relationship between quadrilaterals and parallelograms. Quadrilaterals are closed figures that have four sides. Quadrilaterals are divided into parallelograms and trapezoids. Parallelograms are four-sided figures that have two pairs of parallel sides. A trapezoid has only one pair of parallel sides. A rectangle had four 90° angles; a rhombus has four equal sides. When you have a figure with four right angles and four equal sides; you have a square.

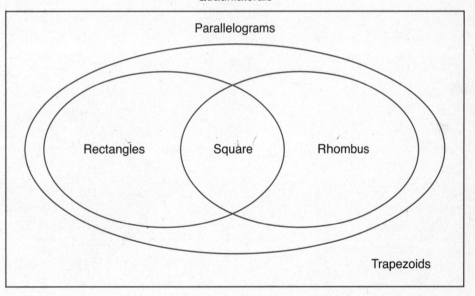

PRACTICE QUESTIONS 3

1. What is the total number of degrees for all the angles in the following figure? THINK!

A. 90°

B. 180°

C. 360°

D. 540°

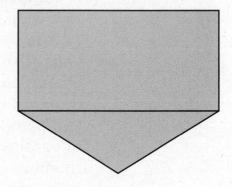

2. Name all the figures you see in the figure below.

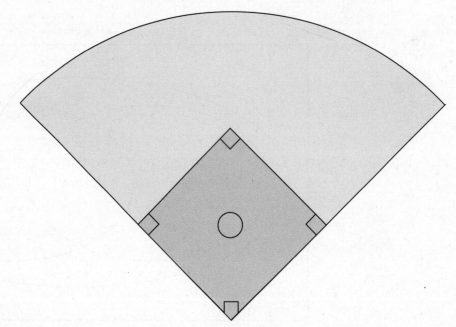

3. Give the definition of a regular polygon.

(Answers on p. 173.)

You now have the building blocks needed to tackle some of the possible questions on the New York State Grade 5 exam.

On the exam, you must calculate the perimeter of regular and irregular polygons. The word perimeter (*peri,* meaning "around," and *meter,* meaning "measure") is the total measure of a figure around its line segments. At this point, we will look at two-dimensional figures.

One particular question you may be asked to solve on the New York State test could be the following.

Aunt Nancy wants to figure out the perimeter of her garden in order to put up a fence to keep the deer out. If her garden has the shape of a regular pentagon and looks like the figure below, how many feet of fence will she need?

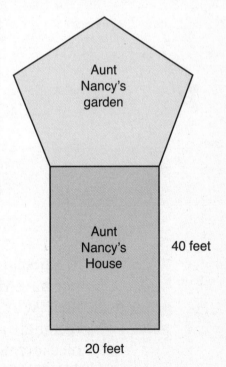

The pentagon is regular. Remember what that means? It means that all of its sides are the same length. We could find out the length of one of its sides and then multiply it by 5 because it is a pentagon. We could also add all the sides together. Since Aunt Nancy's house is in the shape of a rectangle, the length of the side touching the pentagon is also the same length, which means it is 20 feet. You can, therefore, find the perimeter either by (a) adding 20 feet 5 times, which is 20 + 20 + 20 + 20 + 20 = 100 ft, or by (b) multiplying: 20 × 5 sides = 100 ft.

Another way of asking you about the perimeter is to show you a figure that has certain sides labeled. New York State expects you to be able to understand the basic knowledge we discussed previously about geometry in order to figure out the measurement of the other sides. For example, on the exam you may be asked to find the perimeter of the figure below. You will be guided as to how you should be thinking about the problem.

The exam may also ask you to explain your answer. (Typically, students must explain answers in Book 2 of the exam.) Copy the diagram onto a separate sheet of paper and think about possible solutions. Blue lines and numbers will be used to guide you.

Looking at the figure below, what is the perimeter of the figure?

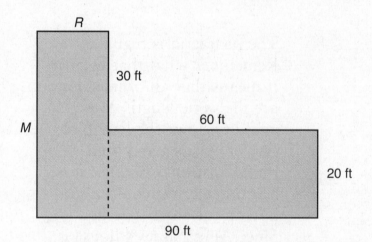

What shape was created with the dotted line that you may not have thought about before? How long is the dotted line? Why? Because the dotted line helped to form a rectangle and it is parallel to the side that is 20 feet, its measurement is also 20 feet. How long is the side marked *M*? Why? Because side *M* is one of the sides of a rectangle and we already know that a rectangle is a parallelogram; it is therefore equal to the sides formed by 30 feet and

20 feet, or 50 feet. Therefore, side M = 50 ft. We now have the following measurements for the figure.

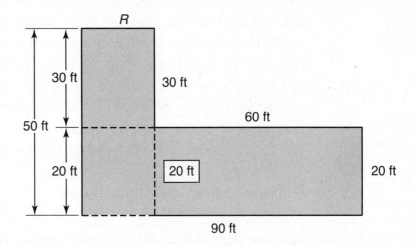

We are almost finished. What is the length of R? Think. R = 90 ft − 60 ft, or 30 ft. Therefore, the perimeter of this figure is 50 ft + 30 ft + 30 ft + 60 ft + 20 ft + 90 ft = 280 ft.

IDENTIFYING PAIRS OF SIMILAR TRIANGLES

This topic deals with the relationship that exists between triangles that have the same angle measurements but different lengths of corresponding sides. Actually, similarity is applicable to other geometric figures as well.

In this figure, both triangles are similar because they have the same shapes but different sizes. All of their *corresponding angles* are equal in measurement and their *corresponding sides* are in a set ratio. We say that $\angle BAC \cong \angle B'A'C'$.

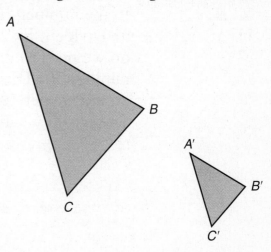

In other words, angle *BAC* is congruent (the symbol we use is ≅) to angle *B'A'C'*. The word **congruent** means "same size and same shape." Therefore, we are saying that the corresponding angles of a similar triangle have the same measurements. We can also name ∠*BAC* as ∠*A* because you can use a single letter to name an angle. The angle used, however, needs to be the letter that corresponds to the vertex of the angle. In other words, when you are referring to ∠*ABC* ≅ ∠*A'B'C'*, it can also be written as ∠*B* ≅ ∠*B'*, where *B* stands for the vertex forming ∠*B*.

When you are describing the ratio of corresponding sides of similar triangles, it can be expressed in several ways. We say that $\overline{AB} \sim \overline{A'B'}$ or that side *AB* is similar to side *A'B'*.

The ratios of the sides of similar triangles are in a set pattern. That is, if we take the sides of △*ABC* and △*A'B'C'*, we see that the following is true: $\dfrac{\overline{AB}}{\overline{BC}} = \dfrac{\overline{A'B'}}{\overline{B'C'}}$.

Using this information, you can find the side *B'C'*.

We can set the ratio as follows: $\dfrac{10}{20} = \dfrac{5}{x}$. We can solve this ratio in two ways.

One way is shown here. What do we need to do to the number 10 to get 5? Divide it by 2. What do we need to do to the number 20? Divide it by 2, and we get 10. Therefore, *x* = 10.

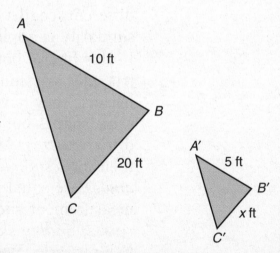

The other way to solve the problem using the multiplication table and a little bit of thinking is as follows. Because we know that this is a ratio, we know that multiplication is involved. So, what number did we multiply to get 2?

In other words, what number did we multiply by 5 to get an answer of 10? That number is 2. And, what number did we multiply the number 2 by to get 20? The answer is 10.

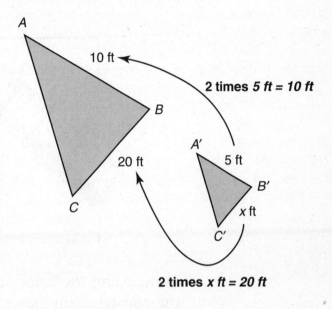

$$\overset{\times}{\frac{10}{20} = \frac{5}{x}}$$

We have almost completed the geometry section. We will present several geometry problems similar to those that you will need to answer on the exam along with their explanations.

As you know, the sum of the angles of a triangle equals 180°. How do we know this? Let us take a look at a triangle.

First, you will need a piece of heavy paper (construction paper is great), scissors, a marker (make sure you do not work on top of your parents' favorite tablecloth or table in case it "bleeds" through the construction paper), and a ruler. Here are the materials needed:

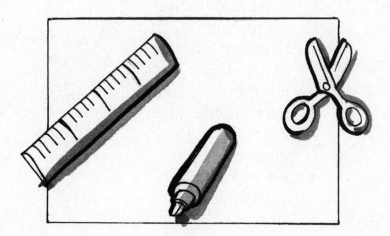

Now, using the ruler and the marker, draw a triangle on the construction paper. Give each angle a letter; you may want to use your initials. Now take the scissors and cut the angles with the letters showing. If you cut the angles in a zigzag, it will help you see a little better. Now take your ruler and marker and make a straight line somewhere on the construction paper away from your work. Then take your scissors and cut along the zigzags you made on the angles and place all three angles aside. Using the marker, make a point on the line you made. Do you remember what the measure of the angle of a straight line is? The measure of the angle of a straight line is 180°; so line $\overleftrightarrow{XY}$ = 180°. If we line up each of the vertices of the angles of $\triangle RAM$, we will be able to see if the sum of the angles of a triangle equals 180°. Let's try it.

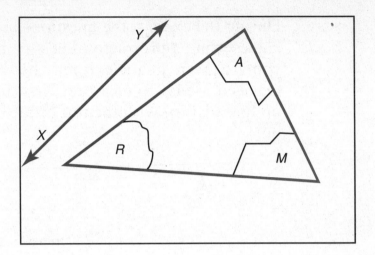

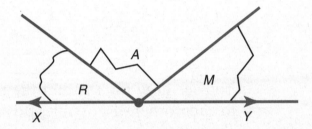

And this is true for any triangle there is in the universe. Last, the sums of the angles of all quadrilaterals equal 360°. This is easy to understand because you know that all the angles of a triangle equal 180°. Let's say we have a trapezoid. At this point, our knowledge of the angles in *any* figure is limited to the angles in triangles. So we can split this trapezoid into two triangles. Let's say we have $\overleftarrow{AB}$, a diagonal that forms two triangles, $\triangle ABC$ and $\triangle ABD$. Since the sum of the angles of each triangle equals 180°, the formula is $2 \times 180° = 360°$. This holds true for all of the quadrilaterals we have studied.

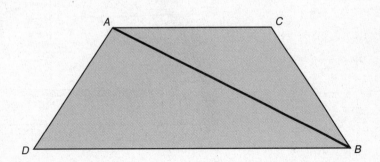

During the exam some questions may ask you to measure angles using a protractor. Let's see how it is done. When you are given an angle, it may be shown in a way that makes it is easy to measure. First, pick the ray that is lined up (parallel to) with the top of the paper you are working on.

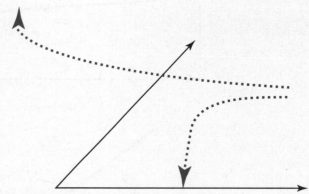

You see how the ray and the top of your paper are parallel to each other. If possible, this is the ray you should choose.

Then examine the protractor. You will notice a few things: One is that it has two sets of numbers, one on top of the arc and the other right underneath. In this case 120° and 60°are such a pair of numbers. For now all you need to know is that if you add both these numbers, you will get 180°.

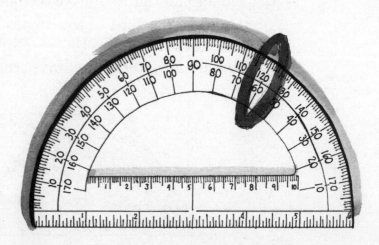

Sometimes, in the middle of the line on the protractor there is a little hole or an arrow, which you should place on the vertex of the angle you are measuring. There is also a line that represents a straight line, and it should be placed over the ray you choose.

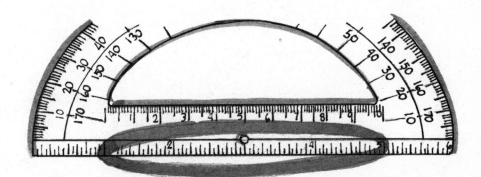

By taking your protractor and aligning it properly, you can measure the degree of any angle. Begin by bringing the protractor's little hole or arrow and the straight line closer to the given angle you need to measure until you

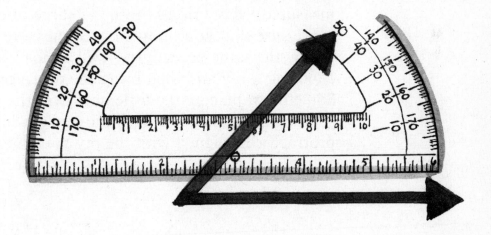

get the top of it to look like the figure below. The line is thicker and extended in order to show you that the angle measured is 50°. The reason for saying that this is a 50° angle is because the angle being measured is an acute angle.

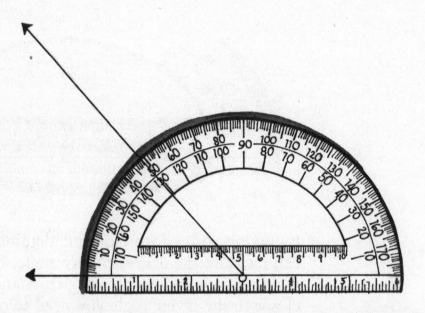

Not all angles are given to you so that they can be easily measured. Many times those in charge of exams want to make sure that you know how to measure the angles. You follow the same procedure—well almost the same. Start by choosing a ray and then bring the protractor toward it. Sometimes, placing the little hole over the vertex and rotating the protractor until the straight line on the protractor is aligned with the ray you chose helps as well.

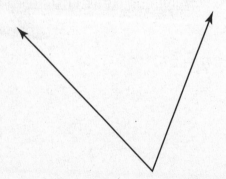

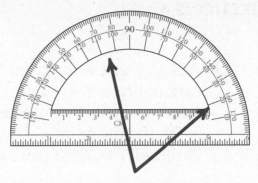

1. First line up the vertex of the angle with the vertex (little hole) of the protractor.

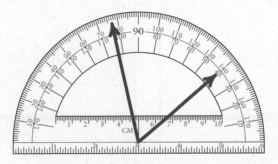

2. Then move the protractor so that one of the horizontal lines on the protractor that intersects the vertex of the protractor lines up with one of the rays of the angle.

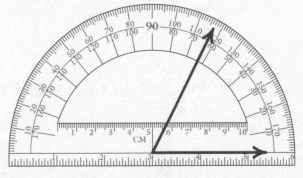

3. Then look at the set of numbers that the other ray crosses. If the angle is acute, use the upper number that appears. If the angle is obtuse, use the lower number that appears. If the angle is a right angle, the top or bottom number will be the same, 90.

PRACTICE QUESTIONS 4

1. A-Rod (Alex Rodriguez) has an enormous baseball card collection. He told Derek Jeter that when placed side by side, his cards reach around his garage. What is the perimeter of A-Rod's garage? (Perimeter: *peri*, "around"; *meter*, "measure.")

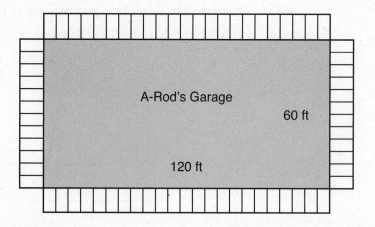

A. 180 feet

B. 7,200 feet

C. 360 feet

D. 320 feet

2. Which of the following is congruent to rhombus *RAME*?

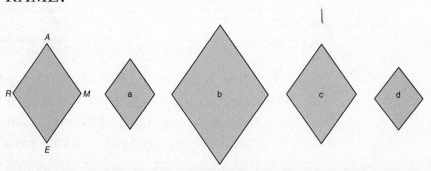

3. Measure each angle of the following triangle using your protractor. Which of the following lists makes the most sense?

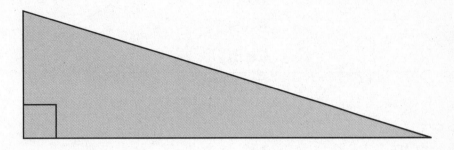

    A. 90°, 90°, 20°

    B. 90°, 10°, 15°

    C. 10°, 30°, 50°

  ⌐ **D.** 90°, 60°, 30°

4. Measure the angle formed from the bottom of the building to the top of the tree. Use your protractor.

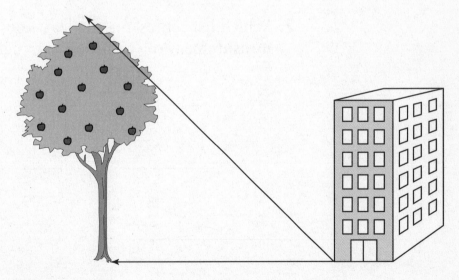

    A. 20°

    B. 110°

  ⌐ **C.** 50°

    D. 90°

5. Measure the following angle.

    A. 60°

    B. 90°

    C. 100°

    D. 145°

6. Measure ∠R.

    A. 100°

    B. 150°

    C. 10°

    D. 40°

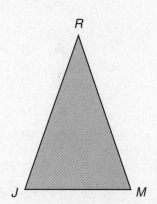

7. Which list represents the least to the greatest in measurement of the following angles?

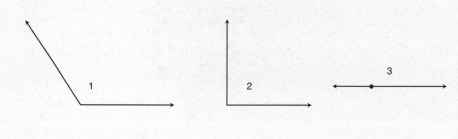

    A. 1, 2, 3, 4

    B. 4, 3, 2, 1

    C. 4, 2, 1, 3

    D. 2, 1, 3, 4

8. Find the measure of ∠C in the right triangle △ABC when the measure of ∠B = 15°.

A. 15°

B. 90°

C. 75°

D. 180°

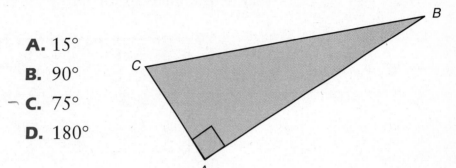

9. The following two triangles are similar. Which of the following fractions represents the ratio between the sides of the small triangle and the sides of the large triangle?

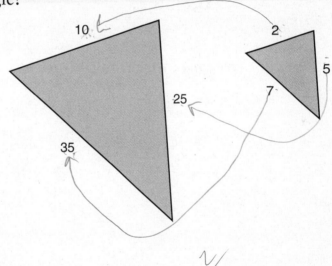

A. $\dfrac{1}{5}$

B. $\dfrac{2}{5}$

C. $\dfrac{2}{7}$

D. $\dfrac{7}{25}$

10. Assume that $\angle D$ and $\angle C$ each measure 90°, $\angle B$ is the only obtuse angle in the quadrilateral, and $\angle A$ = 50°. What is the measure of $\angle B$?

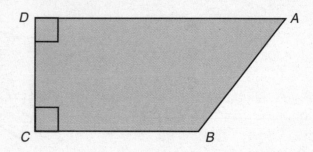

A. 180°

B. 90°

C. 75°

D. 130°

(Answers on pp. 173–176.)

Chapter 3

NUMBER SENSE AND OPERATIONS

In every subject area—mathematics, science, social studies—you need to know certain basics about that subject. In this section you will review the skills needed to have a good understanding of the number systems you have studied or will study in the fifth grade. We will review some of the knowledge you have come across during the past 6 years. First, you need to know how to read numbers. Let's go over some of items that may appear on the exam in March.

READ AND WRITE WHOLE NUMBERS TO MILLIONS

The structure of our numerical system is based on symbols—digits—put together to represent numbers. Sometimes these numbers are part of a system we call whole numbers. For instance, take the following number:

1,492

It is a whole number written in standard form—a form that everyone is familiar with. It has four digits: 1, 4, 9, and 2. Not all digits in a number are created equal. This is important in assigning a **place value** to each digit. Ask yourself, Would I rather have $10.00 or $100.00? Therefore, place value has a place in our study of numbers. In the numeral 1,492 each digit has its place.

They are

Thousands	Hundreds	Tens	Ones
1	4	9	2

Because each digit has a place value, we can say that at least in this case the digit 1 is "worth" 1,000. The digit 4 is worth 400. The digit 9 is worth 90. And the digit 2 is worth 2. In the expanded form numbers are displayed with their digits' values, for example, 1,492.

$$1 \times 1,000 = 1,000$$

$$4 \times 100 = 400$$

$$9 \times 10 = 90$$

$$2 \times 1 = 2$$

or

$$(1 \times 1000) + (4 \times 100) + (9 \times 10) + (2 \times 1)$$

You will learn later this year, or possibly next year, that $n^0 = 1$ when n is equal to any whole number except 0. In other words, $1^0 = 1$ and $6^0 = 1$, and this is true for any whole number except zero.

PRACTICE QUESTIONS 1

1. What is the place value for the underlined digit?

    A. 46,<u>8</u>34   (think—8 is in the hundreds place.)

    B. 67<u>2</u>

    C. 7<u>6</u>8,045

2. Write each of the above numbers in expanded form.

(Answers on pp. 176–177.)

So you will need to know and understand how to work with numbers up to and including 1,000,000. You could learn more place values, but you will be responsible only for those up to and including 1×10^6. We divide numbers into periods in order to read them. Here is a chart that may help.

Millions Period			Thousands Period			Ones Period		
Hundred millions	Ten millions	Millions	Hundred thousands	Ten thousands	Thousands	Hundreds	Tens	Ones

What if you had 34,801 baseball cards and the biggest card dealer in the country wanted to buy them? How would you tell the dealer over the phone how many cards you have?

Hundred millions	Ten millions	Millions	Hundred thousands	Ten thousands	Thousands	Hundreds	Tens	Ones
				3	4	8	0	1

Thirty-four thousand, eight hundred one. Remember that though the zero is in the tens place, you do not say it but still use it to hold a place. Here is the expanded form:

$$30,000 + 4,000 + 800 + 0 + 1$$

Part of how we make sense of numbers and try to understand and study them is to assign an order to each one in order to compare numbers with one another. If you take the previous example of baseball cards, you will understand why that if I have 34,851, it means that I have more cards than you.

	Ten Thousands		Thousands		Hundreds		Tens		Ones
Your cards	30,000	+	4,000	+	800	+	0	+	1
My cards	30,000	+	4,000	+	800	+	50	+	1
	Same		Same		Same		More me		Same

So, I have more cards. Ordering numbers happens in a manner of seconds in our precious gray matter—the brain.

So we say that 34,801 < 34,851; thirty-four thousand, eight hundred one is *less than* thirty-four thousand, eight hundred fifty-one.

Did you know that each July 4th there is an event called Nathan's Hot Dog Competition in New York City where the participants eat as many hot dogs as they possibly can? In 2006, Takeru Kobayashi of Japan ate 53 hot dogs and buns, for a world record. The next year he had stiff competition. On June 2, 2007, Joey "Jaws" Chestnut broke Kobayashi's record with 59 hot dogs and buns in a

qualifying round for the annual Nathan's contest.
Nathan's competition will be an interesting one to follow.
How would you show which number is greater on the
number line?

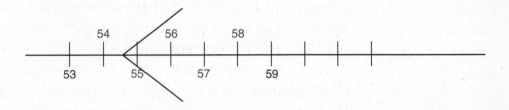

Sometimes when you cannot see the order of numbers,
placing them on a number line helps. Remember that
number lines are created in such a way that the numbers
are arranged from least to greatest—from left to right.
Keep in mind that numbers can also be negative—to the
left of the number zero.

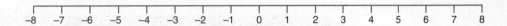

PRACTICE QUESTIONS 2

1. Using the above number line, give two numbers that
are *greater than* 4.

2. Using the above number line, give one number that is
less than 7.

3. Use < or > to complete the following.

A. 6 ___ 4

B. 3 ___ 4

C. 7 ___ 8

D. 8 ___ 7

(Answers on p. 177.)

FRACTIONS

We use the term **fraction** to mean a number that can be written in the form $\frac{a}{b}$, where both a and b can be any integer (an integer is a positive or negative whole number including zero). The number on top, represented by a, is called the **numerator** and tells you how many equal parts of the whole there are. The bottom number, the **denominator**, tells you how many total equal parts there are. A denominator can never be zero. You have seen fractions pretty much all your life—fractions like $\frac{1}{2}, \frac{1}{3}, \frac{2}{5}, \frac{3}{4}$. If you have a block of wood and want to divide it into equal sections, it may look something like this.

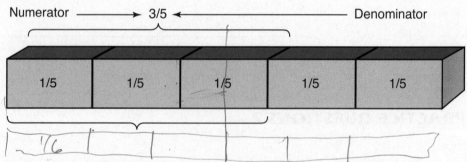

Notice that the denominator lets us know how many pieces to cut the block into, while the numerator tell us how many of these pieces we are interested in.

Try it. For each of the following, first determine the denominator. Then count the section(s) displayed as the numerator and write the fraction represented.

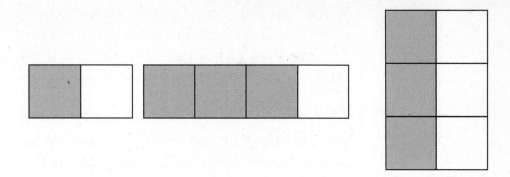

The concept or idea we want to begin to address for most of you, or to review for some, is that of ordering fractions. Let's go back to the number line to assist us in our study. Remember that fractions, or **fractional numbers**, "hang out" between whole numbers; therefore, you will find fractions between two consecutive whole numbers.

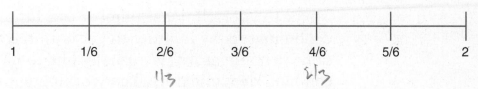

The above represents a number line divided into 6 pieces of equal length between the whole numbers 1 and 2.

COMPARE AND ORDER FRACTIONS WITH UNLIKE DENOMINATORS

Just like the comparison of whole numbers, one of the skills you will need to bring to the exam is comparing fractions. You have to look at two types of fractions that must be compared. The first type are those with the same denominator, for example, $\frac{3}{12}$, $\frac{5}{12}$, $\frac{2}{12}$. All have the same denominator, so they all have the same size portion, but because the numerator is different, it determines which of the fractions is the biggest and which is the smallest.

$$\frac{2}{12} \; < \; \frac{3}{12} \; < \; \frac{5}{12}$$

When determining—figuring out—the order in which to place fractions with different denominators, we know that the sizes of the pieces (denominators) are not the same and therefore that we will not be able to compare them appropriately. So let's say we have $\frac{3}{4}$ and $\frac{1}{5}$ and we want to compare them. Because we have different denominators, we cannot compare them; we need to find a common denominator.

Remember how to do that?

$$\frac{3 \times 5}{4 \times 5} = \frac{15}{20} \qquad \frac{1 \times 4}{5 \times 4} = \frac{4}{20}$$

Think to yourself, what number can I multiply the given denominators by in order to get both fractions to have the same denominator? (We usually like to have the least common denominator when working with fractions.) Remember that what you do to the bottom number you must do to the top number as well. In this case we can make the denominator equal to 20 because 20 is a common multiple of 4 and 5. Let's explore this a little more. If you have $\frac{3}{4}$ and $\frac{4}{7}$, a common denominator will need to be found because they do not have the same denominator. Here is another way of thinking to find the answer. Look at the multiplication table since we are thinking about finding common multiples in order to find a least common denominator. Let's look down the columns for both 4 and 7 to see what numbers we can find that are the same. The first common multiple will also be the least common multiple.

x	0	1	2	3	4	5	6	7	8	9	10	11	12
0	0	0	0	0	0	0	0	0	0	0	0	0	0
1	0	1	2	3	4	5	6	7	8	9	10	11	12
2	0	2	4	6	8	10	12	14	16	18	20	22	24
3	0	3	6	9	12	15	18	21	24	27	30	33	36
4	0	4	8	12	16	20	24	28	32	36	40	44	48
5	0	5	10	15	20	25	30	35	40	45	50	55	60
6	0	6	12	18	24	30	36	42	48	54	60	66	72
7	0	7	14	21	28	35	42	49	56	63	70	77	84
8	0	8	16	24	32	40	48	56	64	72	80	88	96
9	0	9	18	27	36	45	54	63	72	81	90	99	108
10	0	10	20	30	40	50	60	70	80	90	100	110	120
11	0	11	22	33	44	55	66	77	88	99	110	121	132
12	0	12	24	36	48	60	72	84	96	108	120	132	144

Therefore, in this case 28 is the common multiple, and at least in a 12×12 multiplication table, 28 is also the least common multiple. And now the replay: Since we wanted to compare the fractions $\frac{3}{4}$ and $\frac{4}{7}$, we needed to find the common denominator. We found it to be 28, so the question you have to ask yourself is, what number do I multiply 4 by to get 28? And at the same time, what number do I multiply 7 by to get 28? In this case multiplying each denominator by the other will give us the common denominator. Try the following: $\frac{1}{9}$ and $\frac{2}{3}$.

Use the multiplication table to find three common multiples and find the least common multiple, that is, the first multiple in common.

PRACTICE QUESTIONS 3

1. Order the following fractions from least to greatest:
   $\frac{2}{5}, \frac{4}{7}, \frac{1}{6}, \frac{1}{2}.$

2. Make a drawing that represents the following fractions: $\frac{1}{5}, \frac{5}{6}, \frac{2}{3}.$

3. What are four consecutive common multiples for the denominators of $\frac{3}{4}$ and $\frac{2}{3}$? What is the least common multiple of these denominators?

4. Compare the following fractions:

A. $\dfrac{1}{8}$ —— $\dfrac{1}{2}$

B. $\dfrac{2}{3}$ —— $\dfrac{6}{9}$

C. $\dfrac{1}{3}$ —— $\dfrac{4}{5}$

(Answers on p. 177.)

At times, dealing with fractions is not always the best way to work with numbers. Sometimes decimals are the way to go when we want to work with parts of a whole. Like knowing the multiplication table, you should memorize several fractions and their decimal equivalents. Make up flash cards to help you memorize these equivalents.

$$\frac{1}{2} = 0.5 \qquad \frac{1}{7} = 0.142$$

$$\frac{1}{3} = 0.33\overline{3} \qquad \frac{1}{8} = 0.125$$

$$\frac{1}{4} = 0.25 \qquad \frac{1}{9} = 0.11$$

$$\frac{1}{5} = 0.20 \qquad \frac{1}{10} = 0.1$$

$$\frac{1}{6} = 0.167$$

$$
\begin{array}{r}
0.75 \\
4\overline{)30} \\
-28 \\
\hline
20 \\
-20 \\
\end{array}
$$

We can represent all other fractions using this chart. How would you represent $\frac{3}{4}$ in decimal form? Divide 3 by 4. Remember how it is done? You can do this for every fraction you are asked to convert to a decimal.

Just like the digits in whole numbers, the digits in decimals have place values; mathematicians have worked with decimals throughout the centuries.

Hundreds	Tens	Ones	Tenths	Hundredths	Thousandths	Ten-thousandths	Hundred-thousandths	One-millionths
			0.1	0.01	0.001	0.0001	0.00001	0.000001

Just like whole numbers and fractions, decimals can be ordered, and it all has to do with the digits with a particular place value. So, let's say that 34.764 kilograms is the mass of a Sponge Bob Square Pants stuffed animal and the plane that he is traveling on needs to know that weight. How would you read the previous number?

Hundreds	Tens	Ones	Tenths	Hundredths	Thousandths	Ten-thousandths	Hundred-thousandths	One-millionths
			0.1	0.01	0.001	0.0001	0.00001	0.000001
	3	4 • 7	6	4				

First, place each digit into its proper *place value* then read it as if you were reading a whole number. Thirty four and *(after the decimal point, you read it and read the last place value with a digit)* seven hundred sixty-four thousandths— if you had the following: 34.7641.

Hundreds	Tens	Ones	Tenths 0.1	Hundredths 0.01	Thousandths 0.001	Ten-thousandths 0.0001	Hundred-thousandths 0.00001	One-millionths 0.000001
	3	4	7	6	4	1		

Again place each digit in its proper place value and it would read: thirty-four and seven thousand six hundred forty-one ten-thousandths.

PRACTICE QUESTION 4

Use a chart of your own to give the following decimals their proper place values: 145.893, 2.02, 0.076.

(Answer on p. 178.)

Reading decimals and writing them can be a great game between either you and your parents or you and your older brother or sister. Challenge each other in reading and writing decimals. Your textbook and your teacher can give you some work on reading and writing decimals. Once you have become the neighborhood champion in decimal reading and writing, it's time to learn how to put them in order.

Let's start with the decimals you deal with every day. Our money system can help you understand decimals. You know the whole number. Which would you rather have, $1 or $100? With decimals we also have to consider digits and place values. If we have 0.1 and 0.01, which one would you rather have?

First, let us read the numbers. We read: one-tenth and one-hundredth, respectively (in that order). If we are not careful, we might confuse one hundred with one-hundredth. What is the difference? Remember that, like fractions, decimals are parts of a whole. So if you take a whole and divide it into 100 pieces, each piece will be smaller than if you take the same whole and divide into 10 pieces. Therefore, one-tenth is bigger that one-hundredth. Try it. Of the following numbers, which is the smallest and which is the biggest?

0.06, 0.04, 0.045, 0.55, 0.006

Hundreds	Tens	Ones	Tenths 0.1	Hundredths 0.01	Thousandths 0.001	Ten-thousandths 0.0001	Hundred-thousandths 0.00001	One-millionths 0.000001
			0	6				
			0	4				
			0	4	5			
			5	5				
			0	0	6			

Until you can think through this, you may want to use the chart. Why is 0.55 the greatest? Because in the tenths place, the values for the other four decimals are 0 compared to 5 for 0.55. It is the only decimal with a nonzero digit in the tenths place.

Since all the other decimals have a zero in the tenths place, they are equal up to that point. It was mentioned before that the place value of a digit in a decimal tells us how many pieces into which we divide a whole. For example, hundredths tells us that we have divided a whole into 100 pieces (pretty small). Each of the digits with a particular place value tells us how many of these pieces we are talking about. In our example, the hundredths place in 0.06 has the digit 6 in it, which means that we have 6 pieces out of the 100. Now we have to consider the hundredths place values for the remaining digits. Only one of the decimals has the digit 6 in it, and since 6 > 4, 0.06 is the next biggest decimal. Up to now we have the following:

$$0.06 < 0.55$$

Now we have to consider the other three decimals. We have 0.006, 0.04, and 0.045. Which one do *you* think is the smallest? Look at the hundredths place. Both 0.04 and 0.045 have the digit 4 in the hundredths place, and 0.006 has the digit 0 in the hundredths place. Therefore,

$$0.006 < \quad < \quad < 0.06 < 0.55$$

Now for the hardest part—which one of the remaining decimals is smaller, 0.045 or 0.04? Both decimals have the digit 4 in the hundredths place. So let's look at the digit to the right in the thousandths place: 0.045 has the digit 5, and 0.04 has the digit 0 because when there is no number to the right, the place value is always 0.

Therefore, 0.04 can be written 0.040. So,

$$0.006 < 0.04 < 0.45 < 0.06 < 0.55$$

The word **percent** (*per*, "each"; *cent*, "hundred") means "part of 100." We can represent percent in both decimals and fractions most of the time, though we usually represent it in decimals. So let's say that you are playing baseball and you are being coached by your favorite player, and he has a batting average of 0.340 for the year. What percent of the time was he on base? Well it's funny that you should ask that question. For the reason mentioned previously (percent means "part of 100"), we can find that out by multiplying the given decimal by 100, the result we get will be in terms of what part of 100 the given decimal is. Therefore, if your favorite player has a 0.340 batting average, the percent of time he was on base is 0.340 × 100 = 34.0%, which may not seem like much, getting on base 34 times out of every 100 times, but in baseball your favorite player is having a great year. Now if we say that you do your homework 0.340 of the time, I do not think your parents would say that this is good enough. So depending on the context in which you deal with percent, 34% may be okay, but in other cases—completing your homework—that percent is not appropriate.

Look at the following. If you want to purchase a Wii system, the retail price (the price the company thinks you should pay for its product) is $249.99, and Target is selling it at 20% off, what is the *price you will pay*? Remember that 20% is 20/100 or 0.20. Before you answer this question, think about the process.

Are you going to pay more or less for the Wii? Less. How can you find out how much? What does percent mean? Part out of 100. How many hundreds do you have here? Two hundred and some change—$49.99 to be exact. Therefore, 20% of 200 is $40.00, so you will get a little more than $40.00 off the Wii system at Target. Let's find out if your estimate is correct.

$$
\begin{array}{r}
249.99 \\
\times\ 0.20 \\
\hline
00000 \\
499980 \\
\end{array}
$$

There are four decimal places; therefore you will need four more 4 spaces for the answer starting on the right.

$49.9980 is your savings. Which is pretty close to what we originally said.

So, the price you will pay is:

$249.99 (original price)
– 49.99 (amount of discount)
$200.00 (price after savings)

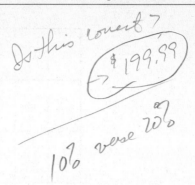

NUMBER THEORY

As you continue with your study of numbers, you will realize that they behave in a certain way. There are two types of numbers: prime numbers and composite numbers. A prime number is a number that is divisible only by 1 and by itself, for example, 2, 3, 5, 7, 11, 13, 17. You may have noticed that it seems that 2 is the only prime number that is also even. It is. But think about it for a moment. A prime number is a number that is divisible only by 1 and by the number itself, and by virtue of what an even number is (divisible by the number 2 without a remainder), a prime number cannot be even, with the obvious exception of the number 2.

You do have to remember that not all odd numbers are prime. For example, 15 is an odd number, but it is not a prime number because not only is it divisible by 1 and 15 but it is also divisible by 3 and 5.

All the rest are composite numbers: each one is divisible by more than 1 and by itself, like 4, 6, 8, 9, 10, 12, 15, 16.

You were previously shown a multiplication table, and in it you will be able to find the multiples of any number. When you are asked for a multiple, you are actually being asked to remember the multiplication table. Do not worry and think you are not smart if you do not know all of your multiplication facts. Many of us adults do not know them; we just have to practice them in order to remember them, and so do you. So, here is a number, 8; find 7 consecutive multiples of 8. Let's look at the multiplication table and find the first 7 multiples of 8.

x	0	1	2	3	4	5	6	7	8	9	10	11	12
0	0	0	0	0	0	0	0	0	0	0	0	0	0
1	0	1	2	3	4	5	6	7	8	9	10	11	12
2	0	2	4	6	8	10	12	14	16	18	20	22	24
3	0	3	6	9	12	15	18	21	24	27	30	33	36
4	0	4	8	12	16	20	24	28	32	36	40	44	48
5	0	5	10	15	20	25	30	35	40	45	50	55	60
6	0	6	12	18	24	30	36	42	48	54	60	66	72
7	0	7	14	21	28	35	42	49	56	63	70	77	84
8	0	8	16	24	32	40	48	56	64	72	80	88	96
9	0	9	18	27	36	45	54	63	72	81	90	99	108
10	0	10	20	30	40	50	60	70	80	90	100	110	120
11	0	11	22	33	44	55	66	77	88	99	110	121	132
12	0	12	24	36	48	60	72	84	96	108	120	132	144

Now look for the least common multiple of 8 and 9. Let's look at what that means: least—smallest; common—same; multiple—multiplication. So in the multiplication table, we look for the multiple that is common to both and is the smallest. Believe it or not, the least common multiple is 72. You try it now. What is the least common multiple for

1. 4 and 7?   28
2. 5 and 3?   15
3. 2 and 9?   18

Knowing the multiplication table will help you find least common multiples. Another fact you will need to know about numbers is that they have factors—numbers that divide a number evenly, that is, without a remainder. The factors of 30 are shown here:

$$30 = 5 \times 6$$

If you are asked for the prime factors, remember what prime numbers are. The prime factors of 30 are shown here.

$$30 = \underset{\substack{\text{Prime} \\ \text{number}}}{5} \times \underset{\substack{\text{Composite} \\ \text{number}}}{6}$$

$$30 = \underset{\substack{\text{Prime} \\ \text{number}}}{5} \times \underset{\substack{\text{Prime} \\ \text{numbers}}}{3 \times 2} \ \} \text{Prime factors}$$

You may be asking yourself, "Why do I need to know about factors?" Well I will tell you why. When working with fractions, you will need to reduce them in order to make problems more manageable. And if you know how to find the greatest common factors of numbers, you will be able to reduce them easier and finish the problem quicker. But remember that acquiring this skill, like anything else in life, takes time. Right now you have enough time to learn the skill, practice the skill, and master the skill. Let me show what I mean about factors in a fraction problem. Knowing you need to reduce a number, you will try to find its factors.

$$\frac{120}{150} \Big\} \quad \frac{12 \times 10}{15 \times 10}$$

$$\frac{120}{150} \Big\} \quad \frac{2 \times 6 \times 10}{3 \times 5 \times 10}$$

$$\frac{120}{150} \Big\} \quad \frac{2 \times 2 \times 3 \times 10}{3 \times 5 \times 10}$$

$$\frac{120}{150} \Big\} \quad \frac{2 \times 2 \times 3 \times 10}{5 \times 3 \times 10}$$

$$\frac{120}{150} \Big\} \quad \frac{2 \times 2 \times 30}{5 \times 30}$$

$$\frac{120}{150} \Big\} \quad \frac{2 \times 2 \times 30}{5 \times 30} \Big\} \quad \frac{120}{150} \Big\} \frac{4}{5}$$

Okay, take a break. Then come back so that we can continue.

PRACTICE QUESTIONS 5

1. Which expression is true?

    A. $\frac{2}{3}$ ⟩ 0.67

    B. $\frac{2}{3}$ ⟩ 0.60

    C. $\frac{1}{4}$ ⟩ 0.67

    D. $\frac{1}{3}$ ⟩ 0.60

2. The speed of Serena Williams' first serve during one of her games was 145.962 miles per hour. What was her speed rounded to the nearest tenths of a mile?

    A. 145.9

    B. 145.96

    C. 146.0

    D. 145.97

3. Your favorite game store has 30% of its store's games dedicated to the PS3. What fraction of the store is dedicated to the PS3?

    A. $\frac{3}{100}$

    B. $\frac{30}{100}$

    C. $\frac{30}{10}$

    D. $\frac{1}{30}$

4. During a recent baseball World Series, 950,000 fifth-grade students watched the entire series. How many 10 thousands are equal to 950,000?

  A. 95

  B. 950

  C. 9,500

  D. 95,000

5. What fraction of these squares represents those that are shaded?

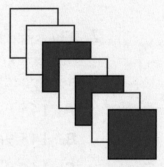

  A. $\dfrac{2}{7}$

  B. $\dfrac{3}{7}$

  C. $\dfrac{7}{2}$

  D. $\dfrac{7}{5}$

6. What fraction of these numbers are prime numbers?

16	2	8	9	7	6	3

A. $\frac{2}{3}$

B. $\frac{3}{7}$

C. $\frac{5}{7}$

D. $\frac{2}{7}$

7. $\frac{2}{4} + \frac{1}{4} = ?$

A. $\frac{3}{8}$

B. $\frac{3}{16}$

C. $\frac{3}{4}$

D. $\frac{2}{3}$

(Answers on pp. 178–179.)

PROBLEM SOLVING

In the past you have probably said to yourself: "I do not understand this problem," "I'm dumb," or "This problem is just too hard for me to do." Yet, at the same time you have encountered and solved problems like this one: I have $10.00, and the game I want costs $39.95. I found it at Target at a 35% discount. How much more money will I need to ask my parents for in order to get that game? Similarly, during the New York State Grade 5 assessment you will encounter (meet) word problems for which you will need strategies that serve as tools for you to solve problems.

PERSEVERANCE—STAYING WITH SOMETHING EVEN IF IT'S HARD AND YOU WANT TO GIVE UP

In order to go on to the next level of Pokeman or Mario Brothers' Tennis, you have to practice. And in order to solve various types of word problems, you have to practice. The strategies you use for particular word problems will become apparent (clearer) the more you solve different types of problems. Therefore, practice needs to be approached with a discriminating (sharp) intellect.

PRACTICE MAKES PERMANENT—
PERFECT PRACTICE MAKES PERFECT

All these experiences have allowed you to gain wisdom (insights) in solving problems. So whether you have had success in getting an answer to a word problem or, on the contrary, after you have made a plan, followed through with that plan, but still failed to come up with the correct answer, you still walk away with something new that you have learned. Many times you learn just as much from failure, if not more, than from being successful. Questions like: Why was my answer wrong? Should I have used another strategy? What am I overlooking in trying to answer this question? allow you to develop ways to solve problems in the future.

When you are practicing how to solve a problem, it may help if you have someone else to discuss the problem with; bounce ideas off each other! Many of us do this naturally when explaining to a friend how to reach the next level of Mario Brothers. The same way you use strategies to find out how to get to the next level in Mario Brothers is the way you can use strategies to solve word problems. Your parents, your teachers, and any adult for that manner when encountering (facing) problems they do not understand or are too difficult for them to solve on their own, often ask another adult for help. They work on the problem together, come up with a possible strategy to solve it, and some times are successful. When they learn that their strategies did not solve the problem, then they go back to the "old drawing board."

STEPS IN SOLVING WORD PROBLEMS

1. *Read the problem.*
2. *Understand the problem.* Do you understand it? No? Read it again *slowly*. Read it out loud. Does it make any sense? Does the problem have charts—tables that you can use to obtain information? Does it have words you do not understand? What key words are in the question? Some of the information will be helpful in finding the answer, while other information will be extraneous (extra, more that you need) to solve the problem. If you are not in the habit of solving word problems, you will struggle in the beginning. But with practice, you will become proficient (very good) at solving word problems.
3. *Develop a plan (strategy).* Now that you understand the problem, you can think about ways to solve it. Remember, there are many ways to approach a particular problem. The strategies you choose will depend on how you understand the information as you read the problem. Some strategies may get you to the answer quicker, while with others it may take a little longer, but they will get you there nonetheless.
4. *Solve the problem.* Just do it! Solve the problem using your chosen strategy(ies). If you can't solve it with your strategy, walk away for few minutes and try something else. Remember, *do not give up*. If you have the time, think about why the strategy you chose did not work; again you will gain wisdom from failure as well as with success.
5. *Look back.* Wow! You did it. Now for the hard question: Does your answer make sense? If it does not, ask yourself why not? You will learn from trying a problem and failing as much as from getting the correct answer.

PRACTICE SOLVING WORD PROBLEMS

Lets begin by trying the following problem. Remember to follow the steps for solving problems. Try this problem on your own. *Do not look ahead.*

In preparation for a jazz concert, Raul, the saxophone player, practiced the following numbers of hours per day for 5 days prior to the concert.

Day Before Concert	Time (in hours)
5	$5\frac{2}{7}$
4	$3\frac{3}{7}$
3	$3\frac{1}{7}$
2	$6\frac{4}{7}$
1	$6\frac{3}{7}$

1. What was the total number of hours that Raul practiced?
2. What was the average number of hours he prepared for the concert each day?

_____

_____

_____

_____

_____

Solve this problem keeping in mind some of the things we spoke of earlier.

In preparation for a jazz concert, Raul, the saxophone player, practiced the following number of hours per day for 5 days prior to the concert.

Day Before Concert	Time (in hours)
5	$5\frac{2}{7}$
4	$3\frac{3}{7}$
3	$3\frac{2}{7}$
2	$6\frac{4}{7}$
1	$6\frac{3}{7}$

> Think!
>
> 1. What information is important?
> 2. What arithmetic skills will I need to solve this problem?
> 3. What strategy will I need?

1. What was the total number of hours that Raul practiced?
2. What was the average number of hours he prepared for the concert each day?

Let's answer the questions using the "think box" as a guide. These questions will help guide us through the problem.

The first question asks for the *total* number of hours Raul practiced. What will we need to do? Add all the hours he practiced during those 5 days. So if we add the hours mathematically, we'll get

$$5\frac{2}{7} + 3\frac{3}{7} + 3\frac{2}{7} + 6\frac{4}{7} + 6\frac{3}{7} = 23\frac{14}{7} \text{ or } 25 \text{ hours}$$

The second question asked you to find the average number of hours Raul practiced a day. Do you remember what the definition of "average" is? Since you have already found the total number of hours he practiced, in order to find the average, divide that amount by the 5 days he practiced: 25 hours divided by 5. Therefore, Raul practiced an average of 5 hours each day.

Okay, take a minute to think about three things you have just learned and write them down.

1. _____

2. _____

3. _____

PRACTICE QUESTION 1

Practice the problem-solving steps.

From October 2004 until January 2005 you saved the following in a money market account (a form of savings) from doing chores, and from birthday and Christmas gifts.

Month	Amount at the End of Each Month (in dollars)
October 2004	27
November 2004	34
December 2004	45
January 2005	68

Use this information (data) to make two graphs (one line graph and one histogram) to show your savings at the end of each month.

Be sure to

- Title your graph.
- Label both axes.
- Graph all the data.
- Scale the graph.

Extra: Additional possible questions you may be asked to answer in Book 2:

1. How would you display the running total per month?
2. At the beginning of February what is the total amount you have contributed (put in) plus the interest your investments have earned?

(Answers on pp. 179–180.)

PROBLEM-SOLVING STRATEGIES

Let us begin developing different types of strategies for solving problems. Remember, some problems can be solved by many of these strategies; it will depend on the following.

1. How you want to approach the problem and which strategy you want to use.
2. Whether or not a specific method is being asked for in the question; for example, the last problem illustrated the use of a graph to show the data.
3. Whether or not you know one or many ways of solving a particular problem. If you know only one of two ways of solving a problem, your parents may have purchased this book to expand your skills in

problem solving. Or it may have been because you just needed a little practice to prepare for the New York State Grade 5 Mathematics Exam.

Here are the problem-solving skills you will need.

USE OR MAKE TABLES OR CHARTS

In Ramon and Mohammed's social studies class, the students were asked to design a community service project. Mohammed and Ramon worked together and came up with the collection of canned goods for those in need at their local city mission. Also, they were able to collect money for the upcoming Thanksgiving season. Each day they collected 23 cans more then the previous day, and they collected $1\frac{1}{2}$ times more money than they had the previous day. Local merchants started them off in week 1 with 56 cans of canned goods and $85.

What amount of canned goods will they collect in 6 weeks?

Think!

1. What is the question to be answered? Place a rectangle around that question.
2. What facts are you going to need to answer that question? Highlight the facts.
3. How can you display the information to help answer the question?

Using a table will allow us to see what is happening.

Week	Number of Cans
1	56
2	56 + 23 = 79
3	79 + 23 = 102
4	102 + 23 = 125
5	125 + 23 = 148
6	148 + 23 = 171

Take a few minutes and walk away. When you come back, think about the following.

1. Could you have done this problem in another way?
2. Have you learned anything from this problem?
3. If so, what?

_____

_____

_____

_____

DRAW A PICTURE OR DIAGRAM

During a NASCAR race, cars can sometimes be too close to each other or separated by only a small distance. The average car is 15 feet in length. If in a particular race there are 12 cars and the organizers of the race have 2 cars per row, how long is the "pack"—all the cars together—if there is 6 feet between each row? Use the same think box questions to start. Let's make a drawing that represents the straightaway of a car track and place our cars. There are 12 cars in this race and 2 cars in each row; therefore, 12 ÷ 2 = 6 rows of cars and there are 6 feet between the rows. Since there are 6 rows and the space between each row is the 6-foot space between each car, there will be 5 such spaces.

Think!

1. What are the questions to be answered? Place a rectangle around each of the questions. You will answer one question at a time.
2. What facts are going to be needed to answer each *question*? Highlight those.
3. How can we display the information to help us answer the question?

So, we then have 6 ft × 5 spaces = 30 ft. Add to this amount 90 ft, the number of feet in the cars, and the result will be 30 ft + 90 ft = 120 ft from the front bumper of the first car to the rear bumper of the last car.

| 15 ft | 6 ft | 15 ft | 6 ft | 15 ft | 6 ft | 15 ft | 6 ft | 15 ft | 6 ft | 15 ft |

| R1 | R2 | R3 | R4 | R5 | R6 |

| 15 ft | 6 ft | 15 ft | 6 ft | 15 ft | 6 ft | 15 ft | 6 ft | 15 ft | 6 ft | 15 ft |

ACT IT OUT

You have a collection of baseball cards, and you have grouped them according to the position of each player. You have 234 cards in total: 26 of the cards are pitchers and 13 are catchers. Of the remaining number of cards that you have, twice as many cards are infielders as outfielders. How many cards are in each group? For this problem, you could use the actual cards or use objects to represent the cards, such as pennies, paper clips, or bingo markers.

Think!

1. What is the question to be answered? Place a rectangle around the question.
2. What facts are needed to answer THAT question? Highlight them.
3. How can you display the information to help answer the question?

Once you have a group of 234 cards (or objects representing the cards), and with the facts from the problem, you can figure out the amount in each category.

Number of Cards	Position
26	Pitchers
13	Catchers
?	Outfielders and infielders

How many cards are in the group of outfielders and infielders? How can you get this amount? You have 234 cards in total, and you know that 13 are catchers and 26 are pitchers. If you subtract the sum of pitchers and catchers from the total number of cards, you are left with 195. In other words, 234 total cards − (13 catchers + 26 pitchers) = 195.

Now, you have 195 cards that have to be divided between the other two groups. So the question now is finding out how you can use the information. Say that the circle ◯ represents the outfielder group, but you do not know at this time how many cards are in this group; it just represents the outfielders. We also know that there are twice as many infielders as there are outfielders; therefore, {◯ ◯} represents the infielders since there are twice as many infielders as outfielders. Again you do not know how many cards are in each group. You know that whatever the number of outfielders is, you can double it in order to find the number of infielders. Also keep in mind that each of the circles representing the outfielders has the same amount of cards inside it. If you add each of the groups together, you will have something like

$$\bigcirc + \{\bigcirc\ \bigcirc\} = 195$$

$$3\ \bigcirc = 195$$

This means that 3 times ◯ equals 195 cards. Divide 195 by 3 to find out how many cards the circle represents. That will be 65. So, if you go back and change each circle to the number 65, you will get 65 + {65 + 65} = 195. Therefore, there are 130 infielder cards and 65 outfielder cards.

The chart looks like

Number of Cards	Position	Number of Cards	Position
26	Pitcher	65	Outfielder
13	Catcher	130	Infielder

WORK BACKWARD

This strategy is usually used when you are dealing with a problem where the answer is known, but the question must be answered at the beginning. For instance, What time did the party start? or At what time did he come into the office this morning? Let us look at a problem.

There is only 1 week before the end of the school year, and you have $7.56 in your pocket on Sunday night. You cannot remember how much money you had on Monday, the beginning of the week. You have receipts, and therefore, you know that you spent $2.59 on a chocolate milk shake at the mall on Wednesday, $15.25 on a T-shirt on Friday, and $30.00 for a video game on Saturday.

What is the question you have to answer? Put a rectangle around it and highlight the facts. Working backward you are able to solve the problem. Since you ended with $7.56, in order to find out how much money you had on Monday, you will need to add what you spent each day working backward.

Sunday Night	Saturday	Friday	Wednesday	Monday
$7.56 to start	$7.56 + $30.00	$7.56 + $30 + $15.25	$7.56 + $30 + $15.25 + $2.59	$55.40
How much money you have	$37.56	$51.81	$55.40	

Here is another one of these types of problems. Using the think box, place a rectangle around the question and highlight the facts.

In order to beat your previous time in the 5K race by 00:01:15, you will have to cross the finish line when the stopwatch indicates 00:21:30. What was your previous time?

In order for you to beat your previous time, you will have to cross the finish line when the stopwatch indicates 00:21:30, and the amount of time you will need to "shave off" is 00:01:15.

> Finish line time: 00:21:30
> Shaved off time: 00:01:15
> Previous best time: 00:21:30 + 00:01:15 = 00:22:45

GUESS AND CHECK

You might say to yourself, "I am stuck. I have read and reread this problem, and I still cannot see a strategy that could be useful." At times when solving a problem, none of the strategies seems to fit. A guess-and-check strategy may be what you need to use. When you use this strategy, make sure you think and reflect on your answer. It should make sense; ask your friends or teachers for their opinions. Write down what you were thinking at the time you encountered this problem so you have a reference point for future discussion and learning when you encounter another problem that is similar.

Michael's Aunt Louise always has many coins in her purse. She tells him if he can guess how many of each denomination (type) of coin she has, she will give them to him. Aunt Louise tells Michael that she has a total of $2.70, that she has 18 coins in total, and that she has only quarters, dimes, and nickels. How many of each does she have?

You know the routine by now: Draw the rectangle and highlight the facts. Let us take a logical guess.

Your guess is (coins must equal 18)	Check	Observations (too high or too low)
10 quarters 4 dimes 4 nickels	$10 \times 0.25 = \$2.50$ $4 \times 0.10 = \$0.40$ $4 \times 0.05 = \$0.20$	$3.40—too high. Fewer quarters, more dimes, and more nickels.
6 quarters 5 dimes 7 nickels	$6 \times 0.25 = \$1.50$ $5 \times 0.10 = \$0.50$ $7 \times 0.05 = \$0.35$	$2.35—too low. More quarters, same numbers of dimes, and fewer nickels.
8 quarters 5 dimes 5 nickels	$8 \times 0.25 = \$2.00$ $5 \times 0.10 = \$0.50$ $5 \times 0.05 = \$0.25$	$2.75—better. Same number of quarters, fewer dimes, and more nickels.
8 quarters 4 dimes 6 nickels	$8 \times 0.25 = \$2.00$ $4 \times 0.10 = \$0.40$ $6 \times 0.05 = \$0.30$	$2.70 Hurray!! Michael is a winner.

Remember the nursery rhyme about Old McDonald and his farm and animals? What if you knew that Old McDonald had some cows and turkeys?

Your class went on a field trip, and while you were distracted you counted 37 heads and 116 feet. You got back to your class and were asked to write a paragraph about your observations for the day. You raised your hand and said, "I noticed that they were 37 heads and 116 feet among the cows and the turkeys." Your friends began thinking about how many cows and turkeys Old McDonald had. How many cows and turkeys are there in Old McDonald's farm?

So you look at the problem and start by drawing a rectangle around the question and highlighting the facts. On Old McDonald's farm each turkey had 2 feet and each cow had 4 feet. Start off by making a guess.

Number of Tries	Number of Cows	Number of Turkeys	Number of Heads	Number of Feet
One	30	7	30 + 7 = 37	30 × 4 = 112 7 × 2 = 14 High = 126
Two	20	17	20 + 17 = 37	20 × 4 = 80 17 × 2 = 34 = 114 Good guess. Try again and think, we need two more feet
Three	21	16	21 + 16 = 37	21 × 4 = 84 16 × 2 = 32 = 116

FIND A PATTERN

This strategy should be used when the problem has patterns that you have to use to get the answer. Let us start with number patterns. Complete the following pattern: 34, 36, 40, 46, 54, ____, ____, ____. This is the way to attack these pattern problems:

1. Find out the difference between two given numbers.

   a. What is the difference between 34 and 36? 36 and 40? 40 and 46? 46 and 56?

   b. Then apply that pattern to the next numbers. In this case you are asked for the next three numbers.

So, $36 - 34 = 2$
        $40 - 36 = 4$
        $46 - 40 = 6$
        $54 - 46 = 8$

Therefore, the pattern changes by adding 2 to the difference between the two previous numbers. In other words,

2   4   6   8   10   12   14
34, 36, 40, 46, 54, 64, 76, 90.

2. If the following pattern continues, how many balls will be in the eighth row and how many balls all together will there be?

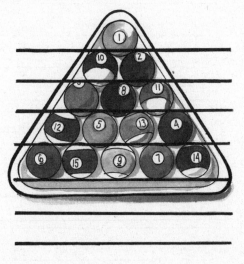

Row 7: _____

Row 8: _____

You will need to answer two questions. The first question is, How many balls will be in the eighth row if this pattern continues? The pattern is that one more ball is added to each row. Therefore, by the time you get to the eighth row you will have 8 balls in that row. You can also think that since you started with 1 ball in row 1, then had 2 balls in row 2, that in the eighth row you will have 8 balls. The second question is to find out how many balls there are all together. There are $8 + 7 + 6 + 5 + 4 + 3 + 2 + 1 = 36$ balls all together.

Reread some of the strategies you just have read. In the next few lines write down some of your observations, things you found useful, and other things that you still need help with and keep track of them.

_____

_____

_____

_____

_____

_____

_____

_____

_____

_____

_____

Here are some questions you may find in Book 2 of the state exam this coming year. Do four of them before taking some time off and doing something else. At the end of this book you will be provided with several more questions in a test-type format. Do your best. You can do this.

PRACTICE QUESTIONS 2

1. Pedro is training for a race. He jogs 4 days a week. Listed below are the distances he ran last week for each day.

PEDRO'S LOG

Day	Distance (in kilometers)
Monday	$5 \frac{3}{4}$
Wednesday	$4 \frac{1}{4}$
Friday	$6 \frac{2}{4}$
Sunday	$4 \frac{3}{4}$

$19 \frac{9}{4} = 21 \frac{1}{4}$
$+ 4 \frac{5}{4}$
$\overline{26}$

Part A: What is the total distance in kilometers that Pedro ran this week?

Show all your work.

_____

_____

_____

_____

_____

_____

_____

_____

_____

_____

_____

_____

_____

_____

_____

_____

Answer: _____

Part B: Janet is Pedro's teammate. If Janet ran 4 $\frac{3}{4}$ kilometers more that Pedro, what was Janet's total distance for the same week?

Show all your work.

_____

_____

_____

_____

_____

_____

_____

_____

_____

_____

_____

_____

_____

_____

_____

Answer: _____

2. Ms. Greenaway, your social studies teacher, has asked your class to keep track of the following stocks on the stock market.

<Stock A>		<Stock B>	
Day	Price per stock (in dollars)	Day	Price per stock (in dollars)
11/13/06	54.89	11/13/06	65.87
11/14/06	52.25	11/14/06	64.78
11/15/06	57.92	11/15/06	65.10
11/16/06	59.85	11/16/06	65.35
11/17/06	62.23	11/17/06	62.90

Use the data to make a bar graph indicating both stock prices at the end of each day the stock market was in session (open).

Be sure to

■ Title the graph.
■ Label both axes.
■ Graph all the data.
■ Provide a scale for the graph.

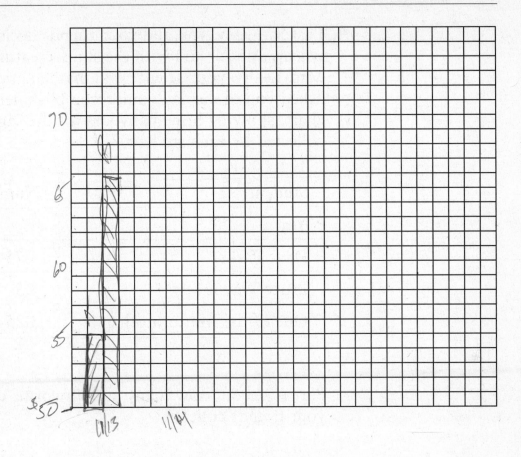

3. Part A: Represent the following decimal in the given grids. 2.34. 1 grid = 1 whole number.

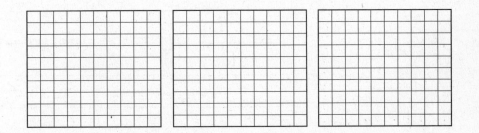

Part B: On the number line below, place the following decimals in order from the least to the greatest.

2.34   2.5   2.16

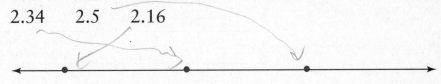

4. Last Saturday, you and your friends decided to collect worms and fill jars with the slimy creatures. Mr. and Mrs. Leahy, your science and mathematics teachers, asked you to display your accomplishments on a chart, showing how many jars (of the same size) were collected by each student.

Student	Number of Jars
You	4.7
Dave Parker	3.75
Bruce Wayne	5.5
One of the Mario brothers	3.25

Part A: About how many jars of worms did you and your friends collect?

Estimate: _____

Explain your answer: _____

_____

_____

_____

_____

_____

_____

_____

Part B: Each jar holds 45 worms. What is your
estimate of the number of worms you and Bruce
Wayne have in your jars?

$45 \times 18 = (45 \times 20) - (45 \times 2)$

$= 900 - 90 = 810$

$11 \times 45 \qquad 495$

$45 \times 18 = (45 \times 20) - (45 \times 2)$

$900 - 90 = 810.$

Estimate: _____

5. Four points on a number line are shown. Place each of the following in its appropriate place above the points.

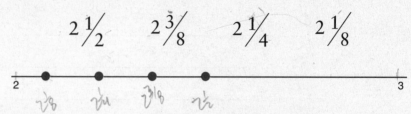

Explain how you arrived at this conclusion.

_____

_____

_____

_____

_____

_____

_____

_____

_____

_____

_____

_____

_____

_____

_____

_____

(Answers on pp. 181–184.)

MEASUREMENTS

THE METRIC SYSTEM

Pick almost any country, and you will have chosen a country that uses the metric system. But if you chose the United States of America (and a few other countries), well that is something different. For the most part the United States does not use the metric system, except when dealing with science—science like physics, chemistry, and engineering. Yet even the United States is slowly coming around.

LENGTH

When we want to use the metric system to measure lengths, we see the it is based on the number 10, for example,

1 centimeter (cm) = 10 millimeters (mm)
1 meter (m) = 100 centimeters (cm)
1 meter (m) = 1,000 millimeters (mm)
1 kilometer (km) = 1,000 meters (m)

Notice that only the number 1 and numbers that can be represented in powers of 10 are used in length measurements. Take a look at these objects and, using a ruler, measure each one.

You will see that the length of the fly is about 6 cm. Using the same ruler, you will see that the iPod length is 10 cm and the width is 6 cm. Did you answer the question about the length of the CD diameter before you knew it was 13 cm? What unit of measure would you use to measure the key shown below?

Centimeter.

What about the penny? Maybe in centimeters, but in millimeters for certain.
What about the bottle cap? Millimeters.
What about the length of a football field? Remember we want to think in the metric system. Meter is correct. Look at the following diagram. It may help you understand the relationship between the different lengths in the metric system.

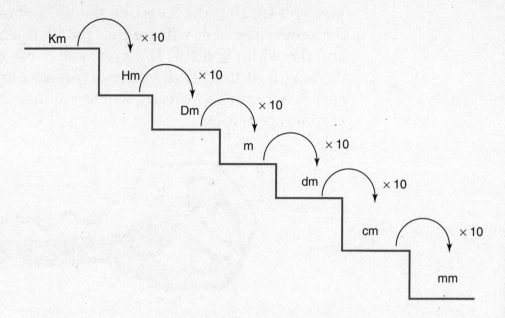

As you come *down* the steps, you *multiply* by powers of 10. As you go *up*, you *divide* by powers of 10. Each step is one power of 10. So if we go from kilometers to centimeters, there are 5 steps. Therefore, you will be *multiplying* by 10^5.

Remember that multiplying by powers of 10 means moving the decimal point to the right. If you have 35 m and you are asked to find out how many millimeters that is, you know that you are going *down* the steps and that you will be *multiplying*.

1. Because you are taking 3 steps, you are going to multiply by 10^3.

2. Then you have

   a. $35 \times 10^3 = 35 \times 10 \times 10 \times 10$

   b. This means you will move the decimal point three places and put zeros in each of the empty places.

   c. $35.0\,0\,0$, which gives us 35,000 mm.

Now if you had 520.1 cm, how many meters would you have?

1. Again you count the steps—in this case just 2, which means 10^2.
2. You are going *up*, which means you will be *dividing*.
3. Therefore, $520.1 \div 10^2 =$

   a. 520.1

   c. 5.201 m

Try these:

1. 8 mm = _____ cm
2. 65 m = _____ mm
3. 38 km = _____ m

The answers are:

1. 0.08 cm   X   0.8
2. 65,000 mm
3. 38,000 m

The other skill you will need for the exam and pretty much for other aspects of your life is converting units of measure within a given measuring system. You have already done this within the metric system, but you will need to learn how to do the same in another system, namely, the system we use here in the United States. Like the metric system, our measuring system, sometimes called the Customary System, has equivalent relationships.

   1 foot (ft) = 12 inches (in.)
   1 yard (yd) = 3 feet (ft.)
   1 mile (mi) = 5,280 ft (ft.)
   1 mile (mi) = 1,760 yards (yd.)

Most of the time you will remember the first two, but if you can memorize the third one, you can figure out the last one. One thing you have to remember in order to convert between two units of measure is the ratio. Yes, the ratio. Let's see how that works. Say you are remodeling a room in the house you live in, and that you know from measuring the outside (the perimeter—the measurement around a figure) of one of the rooms that there are 54 ft around the room. But molding (molding is a piece of wood placed at the bottom of a wall) is sold only by the yard. You want to know how many yards of molding you should buy to finish the room. Therefore, you many want to set the problem up as follows.

1. Let x be the number of yards you will need to go around the room, making sure you round to the nearest ones place.
2. Set up something like this: $x = 54$ ft $\div 3$; $x = 18$ yd.

An additional question could have been asked: "If Lowe's Hardware stores sells molding for $4.75 a yard, how much money would you have to spend to buy the molding? 18 yd $\times$ $4.75 = $85.50. Once again, if you think about the result, you will see that it makes sense that the yards cancel out and you are left with dollars as the unit needed.

TIME

Another type of measuring we all constantly have to deal with involves time. You may have seen timepieces (watches, clocks) on which you can actually see the numbers from 1 through 12, as well as second, minute, and hour hands. These types of clocks and watches are called **analog** clocks and watches.

On other timepieces you may have seen a number say 5 (1–12) hour, colon (:) then another number 34 (0–59) for the minutes and p.m. or a.m., indicating morning or evening. These timepieces are called **digital**.

The following information will be helpful in dealing with the measurement of time.

> 1 minute (min) = 60 seconds (sec)
> 1 hour (hr) = 60 minutes (min)
> 1 day = 24 hours (hr)
> 1 week (wk) = 7 days
> 1 year (yr) = 365 days
> 1 year (yr) = 12 months

Let's say that you and your friends go to the mall to buy some clothing and to go to a movie afterward. Your parents drop you off at 3:45 p.m. and tell you they will meet you in front of the movie theater at 5:20. They also remind you that the movie is a popular one and starts at 5:35. How much time do you have to shop with your friends before meeting your parents at the theater? If you meet your parents on time, how early will you be for the movie?

One way to attack this problem is to count in intervals of 5 minutes from the starting time, 3:45, until the next hour, 4:00. This means that 15 minutes have elapsed (as shown by the distance the minute hand travels between two points on a clock). Then count up to the next hour— in this case 5:00. This means that 1 hour and 15 minutes elapsed between 3:45 and 5:00.

Now you have to do the same thing you did before to count the minutes. Count how many minutes there are between 5:00 and 5:20. There are 20 minutes, so the time that elapses between your being dropped off with your friends and the time you meet your parents is 1 hour and 35 minutes. The other question to be answered is how early will you be for the movie. Count from 5:20 until 5:35 in intervals of 5 minutes, and the answer is 15 minutes.

You are now ready to tackle a few questions on your own.

PRACTICE QUESTIONS 1

1. Find the size of the following item in the units listed below.

    A. cm   *10*

    B. dm   *1*

    C. mm   *100*

2. What metric unit would you use for the following items?

    A. Football field (m, dm, km)

    B. Frame for a picture (m, dm, km)

    C. Size of your foot (m, dm, km)

    D. Distance between your house and the nearest airport (m, dm, km)

3. What is the perimeter of the following figure if you know that the big figure is a rectangle and the small figure is a square?

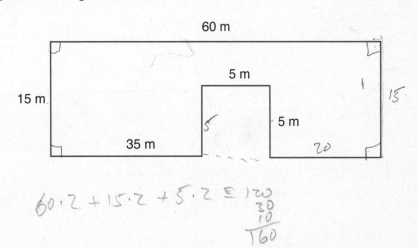

$$60 \cdot 2 + 15 \cdot 2 + 5 \cdot 2 = 120$$
$$\begin{array}{r} 30 \\ 10 \\ \hline 160 \end{array}$$

4. If your middle school basketball team played a game that started at 11:25 a.m. and went into overtime, and ended at 1:26 p.m., how long was the game?

5. Your friend, who lives next to you, said, "Let's go to the shopping mall, which is 260 m away." Does the unit he used make sense? Why? Explain.

6. If I told you that I was 110 years old (actually my grandfather is), how many days would that be?

7. How many centimeters are there in 0.25 m?

(Answers on pp. 185–186.)

Chapter 6

PATTERNS AND FUNCTIONS

T his is a wonderful example of patterns that are available in nature, geometry, and numbers and even in architecture around the country and the world.

You are probably used to seeing the following type of pattern:

$$1, 3, 5, 7, 9, 11$$

or one like this:

The question we will study in this chapter is what is the next number or figure in the pattern. Sometimes what is next will be apparent (easy to see). Other times it will take a little more looking, but it can be done.

Patterns can make us see things we would not otherwise see. What do *you* see in the picture at the top of the page? It will depend on the patterns your eyes are looking at or following and how your brain finds closure to what you eyes are looking at. Show you parents this picture and ask them what they see?

What comes next in each of the following patterns?

1.

2. 30, 2, 40, 2, 50, ___, ___, ___, ___

3. 1, 4, 7, 10, ___, ___, ___

4. 2, 4, 6, ___, ___, ___

5. 6, 30, 150, 750, ___, ___

Answers:

1.

Add a single cube while doubling the number of cylinders.

2. 2, 60, 2, 70 (The digit 2 stays the same; add 10 to the previous non-2 digit.)

3. 13, 16, 19 (Add 3 to the previous number.)

4. 8, 10, 12 (Add 2 to the previous number.)

5. 3,750; 18,750; 93,750 (Multiply the previous number by 5.)

FUNCTIONS

If I am 10 years older than you are, who will be older 6 years from now? You or me? The answer is that I will always be older that you are. This type of relationship is what we call a **function**. A function is like a machine that you tell to do something and it does it every single time without exceptions. In this case, whatever your age is at any point in time, I will always be 10 years older. So, look at it this way:

RULE: +10

Input, Your Age	Output, My Age
1	11
2	12
3	13
4	14
5	15

You can **plot** (graph) a function. A function (rule, input/output machine) is just another way of writing an equation. An **equation** is a mathematical statement consisting of two equal sides separated by an equal (=) sign—hence the word "equation."

Our function is made into an equation by letting y represent your age and m represent my age:

$$y + 10 = m \text{ (read as "}y\text{ plus 10 equals }m\text{")}$$

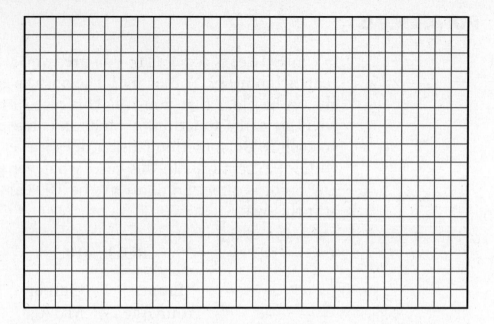

There is a relationship between the input and the output. At least for now, there is only one output for every input.

Look at these relationships and try to figure out the function. What is the rule?

x	1	2	3	4	5	6
y	5	10	15	20	25	30

The rule is to multiply by 5.

What about this one? What is the rule?

Input	9	10	11	12	13	14
Output	89	98	107	116	125	134

The rule is to multiply by 9 and add 8.

VARIABLES AND OPEN MATHEMATICAL SENTENCES

As a mathematician (a person who studies mathematics) you probably have seen something like $3 \times (\ ) = 15$, and as you may know, it is called an **open** sentence. Open because "something" is missing from the sentence for it to be complete. Though it may be simple to figure out, let us use it to find out how to solve the equation. Many times you will see an open sentence in the following format: $3y = 15$, which should be read "Three times y equals fifteen." We call the letter that represents the number we need to find a **variable**. There are many reasons why in some books the variable in a multiplication equation appears right next to the number. If a multiplication sign is included, it may be confused with the variable x.

$3y = 15$ means that there is a number out there in the universe that when multiplied by 3 gives 15. The only possible number is 5. So, when you are asked to solve the equation $3y = 15$, you should use what ever methods you know to solve it and say that $y = 5$. How you solve it is another story. Here is a method I have found to work rather well.

Say you are asked to solve the equation $3y = 15$. Solving an equation is another way of finding the variable.

The diagram shows that the variable y is at the beginning of the equation and is multiplied by 3, the result being 15. Now the question is, How do we get back to what we started with?

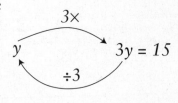

Since on the way to 15 we multiplied by 3, in order to go back, we need to divide by 3. If we divide 15 by 3, the result will be 5.

PRACTICE QUESTIONS 1

1. $16 + n = 15$

2. $5 - w = 0$

3. $9w = 234$

4. $e - 10 = 1,000$

5. $r + 0 = 0$

(Answers on p. 187.)

PRACTICE QUESTIONS 2

1. Your friend buys a Wii controller that costs \$35.95. He has earned \$85.00 from 3 weeks of work. Write an equation that represents how much money he has left.

   $m = 85 - 35.95 = 49.05$

2. My great-grandmother is 5 years older than twice my age. Write an expression that shows the relationship between my age and my great-grandmother's age.

   $G = 5 + 2m$

3. What is the function (rule) that gives the following results?

Input	Output
5	14
6	16
7	18
8	20

$2I + 4 = 0$

4. What is next figure in the following pattern?

5. What are the next two numbers in this pattern?

-3   -4   -5   -6

53, 50, 46, 41, 35, <u> 28 </u>, <u> 2D </u>

6. Find the value of the variable in each of the following equations.

A. $n + 6 = 14$ 8

B. $3x + 5 = 14$ 3

C. $44 - r = 12$ 32

7. Graph the following: $y = 3x + 4$

8. What is next figure in the following pattern?

7 sides

9. Represent the following expressions as equations.

A. I am 1 year older that my brother Luis. $m = L + 1$

B. I am 5 years older than my sister. $m = S + 5$

(Answers on pp. 187–190.)

DATA AND PROBABILITY

Let's start with some definitions.

MEAN

Suppose you want to find the **mean** (the average—also called the **arithmetic average**) of your last 6 years of final grades. In order to find the average you will have to

1. Add your grades.
2. Divide the sum by the number of grades you had.

You had the following grades for the past 6 years:

$$98, 97, 98, 99, 98, 100$$

In order to find the mean you must first add them:

$$98 + 97 + 98 + 99 + 98 + 100 = 590$$

Now you have to divide by the number of grades, which is 6:

$$590 \div 6 = 98.33$$

Therefore, the mean for this set of data is 98.33.

PRACTICE QUESTIONS 1

Find the mean of each of the following.

1. Your favorite basketball player's total points for the first four games:

    35, 45, 55, 63   *49.5*

2. Your speeds in typing class for seven consecutive classes (typing speeds are calculated in words per minute):

    88, 79, 85, 90, 92, 95, 95.   *89.14*

3. The number of hours you have studied this week:

    5, 6, 5, 5, 4, 4, 5   *4.86*

4. Your math scores this year:

    90, 89, 95, 96, 97, 99, 95   *94.4*

5. The amount of money you have saved each month this past school year:

Sept.	Oct.	Nov.	Dec.	Jan.	Feb.	Mar.	April	May	Jun.
$12.25	$52.20	$15.75	$60.25	$10.15	$10.15	$19.68	$16.30	$19.62	$19.63

(Answers on p. 190.)

23.6

MEDIAN

Sometimes when data must be studied a little closer, we need to find the middle number in a set or group of data. We call the middle number the **median**.

The first step in finding the median is to place all the numbers in order. Most of the time the numbers are placed in *ascending* order (least to greatest). The next step is to count the numbers. If the count of numbers is odd, then the median will be the number in the middle. But if the count of numbers is even, then the median will be the mean of the two middle numbers. Here are some examples:

$$12, 52, 45, 15, 62, 53, 70$$

First, place the numbers in order from least to greatest:

$$12, 15, 45, 52, 53, 62, 70$$

If you count the numbers, you will see that there are 7 numbers in this example; therefore, the middle number is the median. There is a box around the median. Notice that there are equal amounts of numbers on either side of 52.

$$12, 15, 45, \boxed{52}, 53, 62, 70.$$

Now if we have

$$100, 95, 98, 90, 96, 97, 92, 99$$

first, we order them:

$$90, 92, 95, \boxed{96, 97}, 98, 99, 100$$

If we take a count of how many numbers there are, we will see that it is 8. So in order for us to find the median, we have to find the mean of the middle two numbers. A square has been placed around both of these numbers. Once again you will see the same amounts of numbers on either side of the two numbers. Now we will find the mean of these numbers:

$$\frac{96 + 97}{2} = 96.5$$

Therefore, the median is 96.5.

PRACTICE QUESTIONS 2

Find the median of each of the following sets of numbers.

1. 5, 2, 10, 8, 4

2. 78, 85, 79, 65, 40

3. 60, 80, 1, 98, 85, 90

4. 767, 747, 787, 737, 777

5. 2, 5, 7, 1, 9, 8

(Answers on pp. 190–191.)

MODE

When a number appears more than any other number within a data example, that particular number is called the mode. Remember though, that if there isn't a number that is the mode, you will need to make that clear by stating that there is no mode. If you use the number zero to indicate there is no mode, because zero IS a number, it will be interpreted that you are stating that zero is the mode of that particular set of data.

PRACTICE QUESTIONS 3

Find the mode of each of the following sets of numbers.

1. 3, 4, 5, 4, 3, 3, 3, 5, 7

2. 10, 12, 12, 5, 5, 8, 1, 6, 12

3. 5, 8, 4, 0, 1, 10, 0

4. 80, 40, 50, 90, 50, 80   none

(Answers on p. 191.)

RANGE

The **range** within a data set is defined as the difference between the largest number and the smallest number the set. You do not necessarily need to order the numbers to find the range, but it is recommended. For example, you have the following data for the temperatures in Buffalo, New York, for the first week of December. What is the temperature range?

Mon.	Tue.	Wed.	Thurs.	Fri.	Sat.	Sun.
27°	23°	31°	45°	21°	45°	23°

21°, 23°, 23°, 27°, 31°, 45°, 45°

Therefore, the range is 45° − 21° = 24°.

PRACTICE QUESTIONS 4

Find the range of each of the following sets of data.

1. 8, 11, 9, 7, 5, 10, 12

2. 89, 79, 50, 65, 98, 10

3. 50, 90, 100, 58, 44, 50

4. 1,300, 1,250, 1,350, 1,200, 1,000, 950

5. 80,000, 85,000, 89,000, 82,000, 78,000, 83,000

(Answers on p. 191.)

INTERPRETATION OF DATA, GRAPHS, AND CHARTS

Have you ever played video games or baseball or football with your friends and it becomes just too confusing? Well at times we cannot handle too much information at one time, or the information is given to us in such a manner that we cannot make "heads or tails" of it. At times like these it is very helpful to sit back and see if displaying or arranging the information in another way would help. For example, let's say that 10 players in the National Football League (NFL) weigh the following (in pounds): player 1, 250; player 2, 214; player 3, 195; player 4, 225; player 5, 235; player 6, 199; player 7, 234; player 8, 304; player 9, 302; player 10, 210. If you are one of the coaches for the NFL team that has this information and needs to make recommendations for stamina (endurance), it may help you determine the type of exercise program you should develop to make your team ready for the season.

The information is correct, but is it shown in a way that you can easily make sense of it? It may not be obvious to some. We could display it in the following way.

A table with the information displayed by a player's last name, first name, or weight could help. Which do you think would be better?

Player	Weight (in pounds)
Player 3 (C)	195
Player 6 (F)	199
Player 10 (J)	210
Player 2 (B)	214
Player 4 (D)	225
Player 7 (G)	234
Player 5 (E)	235
Player 1 (A)	250
Player 9 (I)	302
Player 8 (H)	304

There are a few things that you should notice. First, the weight unit has to be placed in the heading for the reader to understand or know what the numbers represent. Remember, we want a conditioning program for each of these 10 players, but their names do not necessarily have to appear in alphabetical order. We want to know the ranking by order of weight; therefore, the weights are being displayed in ascending order. Now, if we want to place them in alphabetical order (which their numbers now represent), the table would look different. It would look like this; so it truly depends on what was asked for.

Player	Weight (in pounds)
Player 1 (A)	250
Player 2 (B)	214
Player 3 (C)	195
Player 4 (D)	225
Player 5 (E)	235
Player 6 (F)	199
Player 7 (G)	234
Player 8 (H)	304
Player 9 (I)	302
Player 10 (J)	210

At times data that you get can be organized in graphs. If you read the newspaper, you would quickly come across information being displayed in graph format. There are many types of graphs, and at times different graphs can display a particular set of information—data—better than others.

Suppose you want to find out what games students in your class want to play during physical education class and you are given the following choices: football, soccer, tennis, track, baseball, basketball, volleyball, and hockey. How would you find out the information? Once you have a chart, you can take the information and display it in a bar graph.

You could prepare a frequency table like the following.

Sports	Boys	Girls	Total
Football	IIII IIII II	III	15
Soccer	IIII IIII III	IIII IIII	23
Tennis	III	I	4
Track	IIII IIII III	IIII IIII IIII	28
Volleyball	II	I	3
Hockey	I	IIII	5
Baseball	IIII	IIII	8

5th-Grade Sports

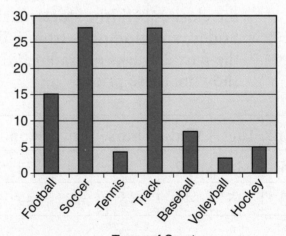

Types of Sport

You could have displayed the gender breakdown as well.

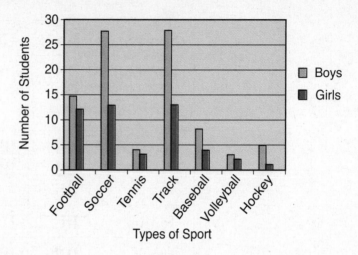

Notice that each of the graphs has a title, that each category is shown at the bottom, and that there is an explanation of the numbers on the side. This allows the reader of the graph to know what he or she is looking at.

There is another type of graph that you often see, a **line graph**. Most of the time you see line graphs when the information being displayed changes over time—for example, information like that the price of stocks on the stock market. If you are keeping track of three stocks at the closing bell (the stock market is open Monday through Friday, and trading is from 9:00 a.m. to 4:00 p.m.) over the past 2 weeks of trading, you can prepare a chart showing these prices.

Stock	Mon.	Tues.	Wed.	Thurs.	Fri.	Mon.	Tues.	Wed.	Thurs.	Fri.
A	35	36	37	25	24	31	35	40	41	45
B	40	45	46	45	47	47	48	47	47	48

You can use a line graph to display this information in order to give your readers a visual ideal without their knowing the specific details—providing a general idea of how each stock is doing.

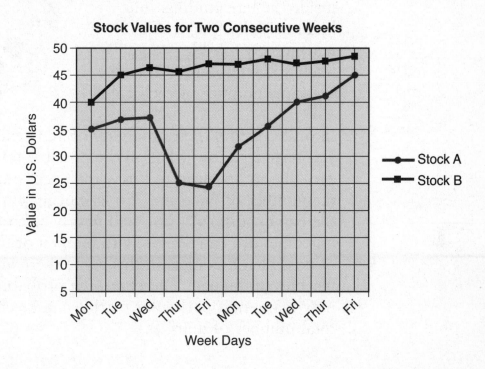

Stock Values for Two Consecutive Weeks

PROBABILITY

Probability is the study of the chance that something will happen. If something or some event is always going to happen, we say it is certain. For example, Tuesday always comes after Monday; therefore, it is certain that Tuesday will follow Monday. If there is no possible way that something will happen we say it is impossible. For example, it is impossible that someone could breathe underwater without a mechanical devise or assistance. An event that is certain has a probability of 1. An event that is impossible has a probability of 0. All other probabilities fall between the certain and the impossible. Events—sometimes called experiments—are likely or unlikely to happen; if an event is just as likely to happen as not to happen, we say that it is equally likely to occur.

If you have a bag with different colored marbles in it and you do not have a little Martian inside the bag guiding your hand to pick a particular marble, any marble is equally likely to be chosen if you have the same number of each color.

In order to find the probability of an event happening, you have to know what all the possible outcomes are. You actually do not need to know the exact outcomes but just the number of outcomes. So let's begin with the marbles in the bag experiment. You will first need to find out what outcomes are equally likely to happen or the total number of possible outcomes. In this game there are 8 total marbles in the bag. Therefore, the probability depends on how many of a particular marble can be chosen out of the total number of marbles.

$$\text{Probability(event)} = \frac{\text{number of times the event occurs}}{\text{total number of possible outcomes}}$$

For example, the probability of a blue marble being chosen will be the number of blue marbles there are in the bag out of the total of 8 marbles:

$$P\text{ (blue)} = \frac{1 \text{ blue marble}}{8 \text{ total marbles}} \text{ or } P\text{ }(b) = \frac{1}{8}$$

$$P\text{ (white)} = \frac{\text{number of white marbles in the bag}}{\text{total number of marbles}} \text{ or } P\text{ }(g) = \frac{4}{8}$$

How we arrange things is important because it will help us find either the number of times—outcomes—or the specific outcome. Say you have a party and you are serving two different cakes, three different ice cream flavors, and four different toppings. What are all the possible outcomes?

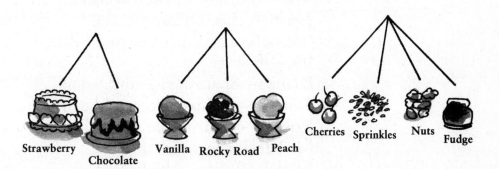

PRACTICE QUESTIONS 5

1. For each of the following sets of data find the

A. Mode

B. Median

C. Mean

D. Range

   i. 52, 19, 60, 40, 52, 16, 55, 60, 78, 60   60/

   ii. 25, 30, 45, 43, 43, 35, 32, 27, 50   43/

   iii. 60, 70, 55, 50, 0, 90, 80   —/

   iv. 33, 20, 21, 16, 21, 5, 19   21/

2. When asked their favorite foods, several students reported the following. Reorganize this information into a tally chart.

Rafael—fish fry
Joshua—Buffalo chicken wings
Tyler—Buffalo chicken wings
Jessica—cheese pizza
Gabrielle—Spanish rice and beans
Lisa—fish fry
Dianna—Spanish rice and beans
Trenton—Spanish rice and beans
Robert—Spanish rice and beans
Michael—cheese pizza
Patrick—Buffalo chicken wings
John—salad
Renee—cheese pizza
Bailey—fish fry
Rick—Spanish hot dogs
Hernando—Spanish rice and beans
Mary—cheese pizza

3. All the students in Ms. Estrella's class participated in the survey and were asked to answer only once. Answer the following questions.

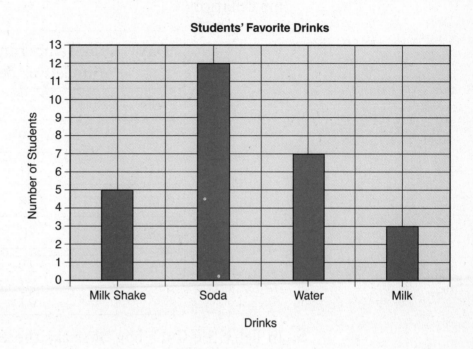

Part A: How many more students said they drank water than drank milk?

_____

_____

_____

_____

Part B: What is the total number of students in Ms. Estrella's class?

_____

_____

_____

_____

4. The average temperatures for the Dominican Republic during the second week of January were the following. Draw a line graph displaying this information.

Day	Temperature (in Celsius degrees)
1	27
2	29
3	31
4	28
5	30
6	29
7	31

5. In Felix the Cat's bag of tricks there are special triangles.

A. List the color and number of each type of triangle.

B. State the total number of triangles.

C. Find *P* (light blue).

D. Find *P* (not light blue).

E. Which triangle has the greatest probability of being chosen and why?

(Answers on pp. 191–193.)

PRACTICE TESTS

TIPS FOR TAKING THE TEST

Here are some suggestions to help you do your best.

1. Be sure to read carefully all the directions in the test book.
2. You may use your tools to help you solve any problem on the test.
3. Read each question carefully and think about the answer before choosing your response.
4. Be sure to show your work when asked. You may receive partial credit if you have shown your work.
5. For Book 1 of the exam you will have 45 minutes to complete the given questions. Use a watch to keep track of time. For this part of the exam calculators are not allowed.
6. For Book 2 of the exam you will have 50 minutes to complete the questions. Use a watch to keep track of time. For this part of the exam calculators are allowed.

7. When you see a picture of a ruler, you may use it to help you solve the problem.

8. When you see a picture of a protractor, you may use it to help you solve the problem

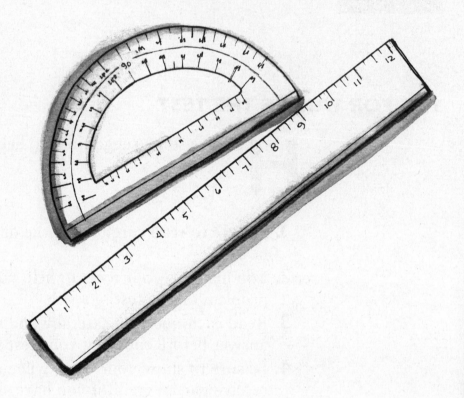

ANSWER SHEET:
PRACTICE TEST 1

Book 1

1. Ⓐ Ⓑ Ⓒ Ⓓ   10. Ⓐ Ⓑ Ⓒ Ⓓ   19. Ⓐ Ⓑ Ⓒ Ⓓ

2. Ⓐ Ⓑ Ⓒ Ⓓ   11. Ⓐ Ⓑ Ⓒ Ⓓ   20. Ⓐ Ⓑ Ⓒ Ⓓ

3. Ⓐ Ⓑ Ⓒ Ⓓ   12. Ⓐ Ⓑ Ⓒ Ⓓ   21. Ⓐ Ⓑ Ⓒ Ⓓ

4. Ⓐ Ⓑ Ⓒ Ⓓ   13. Ⓐ Ⓑ Ⓒ Ⓓ   22. Ⓐ Ⓑ Ⓒ Ⓓ

5. Ⓐ Ⓑ Ⓒ Ⓓ   14. Ⓐ Ⓑ Ⓒ Ⓓ   23. Ⓐ Ⓑ Ⓒ Ⓓ

6. Ⓐ Ⓑ Ⓒ Ⓓ   15. Ⓐ Ⓑ Ⓒ Ⓓ   24. Ⓐ Ⓑ Ⓒ Ⓓ

7. Ⓐ Ⓑ Ⓒ Ⓓ   16. Ⓐ Ⓑ Ⓒ Ⓓ   25. Ⓐ Ⓑ Ⓒ Ⓓ

8. Ⓐ Ⓑ Ⓒ Ⓓ   17. Ⓐ Ⓑ Ⓒ Ⓓ   26. Ⓐ Ⓑ Ⓒ Ⓓ

9. Ⓐ Ⓑ Ⓒ Ⓓ   18. Ⓐ Ⓑ Ⓒ Ⓓ

Book 2   (27–34)

Use the answer spaces in the test booklet to provide your answers and show your work.

PRACTICE TEST 1

Book 1: 45 minutes

Directions: Choose the best answer and fill in the corresponding bubble on the answer sheet.

1. In 2006 the population of Pokeman was 3,008,004. What is the place value of the 8 in this number?

    A. 80,000

    B. 8,000

    C. 800,000

    D. 800

2. Use the ruler to measure the iPod. What is its width?

    A. $2\frac{1}{8}$ inches

    B. $2\frac{1}{3}$ inches

    C. $2\frac{1}{2}$ inches

    D. $2\frac{1}{4}$ inches

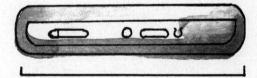

Go On

3. If Air Force 1 used 45% of its fuel, how much fuel did Air Force 1 use?

A. 0.045

B. $\frac{4}{10}$

C. $\frac{45}{100}$

D. 0.5

4. What is the measurement of the third angle of a triangle if one angle is 30° and the second is 20°?

A. 40°

B. 90°

C. 130°

D. 180°

5. In Ms. Morquecho's college class the ratio of male students to female students was 2:5. How would this ratio be written as a fraction?

A. $\frac{5}{2}$

B. $\frac{3}{5}$

C. $\frac{5}{7}$

D. $\frac{2}{5}$

6. The latest hybrid cars are built to accelerate quickly using gas and then run steadily on batteries for miles. Which graph represents this action?

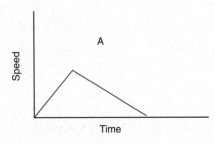

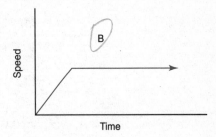

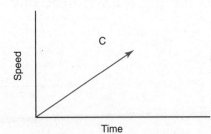

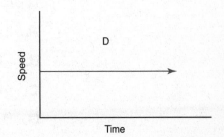

A. graph A

B. graph B

C. graph C

D. graph D

7. If there are 6 marbles in each bag and there are 72 marbles all together, which of the following shows how to find the number of bags?

A. $72 \div 6 = ?$

B. $72 - 6 = ?$

C. $6 \times 72 = ?$

D. $72 + 6 = ?$

Go On

8. You want to buy some Matchbox cars, and each costs $1.20. If you have $5.00, how many Matchbox cars can you buy?

   A. two

   B. three

   C. four

   D. five

9. How much would the value of 8,120 change if we replaced the digit 1 with the digit 7?

   A. 100

   B. 700

   C. 600

   D. 0

10. A hummingbird can flap its wings 120 times every second. How many times can the hummingbird flap its wings in 5 seconds?

    A. 125 times

    B. 500 times

    C. 600 times

    D. 1,000,000 times

11. Joshua, Tyler, and Jessica's father is taking a picture of them to hang on the wall. What amount of framing does he need if the picture has the following dimensions?

    A. 32 inches

    B. 64 inches

    C. 220 inches

    D. 12 inches

12. The fundraiser at your school raised the following amount of money.

MONEY FROM FUNDRAISER

Week	Money (in dollars)
1	250
2	500
3	750
4	???

If this continues, how much money will your school raise during the fourth week?

A. 14,000

B. 1,000

C. 900

D. 1,400

13. In order to run in the Boston Marathon, runners usually spend many hours in daily training. One runner ran from 5:15 a.m. until 7:45 a.m. How many hours did he train?

A. 1 hour, 45 minutes

B. 2 hours

C. 2 hours, 30 minutes

D. 2 hours, 45 minutes

Go On

14. Which measurement makes more sense for the length of an average front yard in a city?

A. 5 inches

B. 5 kilometers

C. 15 meters

D. 55 yards

15. The following chart shows how many home runs Mr. Mercado hit in 4 days.

Day	Number of Home Runs
1	● ●
2	● ● ●
3	● ● ●
4	● ●

If ● represents 2 home runs, how many home runs did Mr. Mercado hit in all 4 days?

A. 10 home runs

B. 20 home runs

C. 30 home runs

D. 2 home runs

16. What is the least common multiple (LCM) of 5 and 4?

A. 1

B. 20

C. 40

D. 80

17. If the following represents the height of a building and the height of a tree, what fraction represents the ratio of the height of the tree to the height of the building?

A. $\frac{65}{10}$

B. $\frac{1}{5}$

C. $\frac{10}{65}$

D. $\frac{6}{5}$

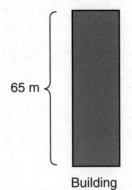
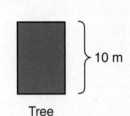

65 m

10 m

Building   Tree

18. If during a jumping rope competition, the average jumper jumps 16 times every 10 seconds, how many jumps does the average jumper jump in 60 seconds?

A. 60

B. 90

C. 66

D. 96

19. What number would make the following open sentence true?

$$34 + (\ ) > 51$$

A. 18

B. 16

C. 17

D. 15

Go On

20. The number of cars going through a particular intersection was counted for six consecutive seconds and the results were

    25, 45, 80, 32, 55, 60

    What is the range for this set of data?

    A. 80

    B. 25

    C. 57.5

    D. 55

21. What is the probability of pressing a number that is a multiple of 3 if there are 12 numbers on a remote control?

    A. $\dfrac{3}{12}$

    B. $\dfrac{4}{7}$

    C. $\dfrac{4}{12}$

    D. $\dfrac{9}{12}$

22. What would be a good estimate of the height of an adult?

    A. 2 meters

    B. 20 meters

    C. 7 meters

    D. 10 meters

23. Which statement is correct?

A. $4 \times (2 + 5) = (4 \times 2) + 5$

B. $4 \times (2 + 5) = (4 \times 2) \times 5$

C. $4 \times (2 + 5) = (4 \times 2) + (4 \times 5)$

D. $4 \times (2 + 5) = (4 + 2) \times (4 + 5)$

24. The following grid represents what decimal number?

A. 1.0

B. 0.08

C. 0.8

D. 1.08

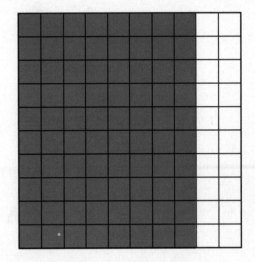

25. What is the area of the following figure?

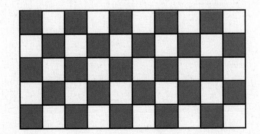

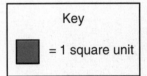

Key

= 1 square unit

A. 5

B. 10

C. 50

D. 30

Go On

26. During art class one of the students, Jessica, drew a set of geometric figures in the following pattern. What figure should come next?

A.

B.

C.

D.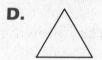

Stop

Book 2: 50 minutes

Directions: Fill in the answers below and show your work.

27. If there are 7 games you want to buy from the game store and each one of them costs $6.95, estimate how much would you have to pay for them (taxes not included).

Answer: $ 49.⁻

Explain how you arrived at your answer.

_____

_____

_____

_____

_____

_____

_____

_____

_____

Go On

28. Triangle *ARM* has the following angle measurements.

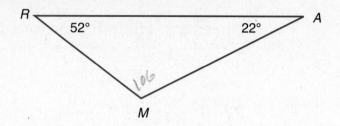

What is the measure of ∠*RMA*?

Answer: _____106_____

Explain how you determined the measure of the angle without using a protractor.

_____

_____

_____

_____

_____

_____

_____

_____

_____

_____

_____

29. Use the following information about the acceleration of sports cars to draw a line graph.

0–60 MPH COMPETITION

Type of Car	Acura NSX	BMW Roadster 3.0 si	Porsche 911 Carrera S	Ferrari F430 Challenger	Lamborghini Callardo
Time (in seconds)	3.5	5.6	4.7	3.3	3.8

Make sure to:
Title the graph.
Label both axes.
Graph all the data.
Provide a scale for the graph.

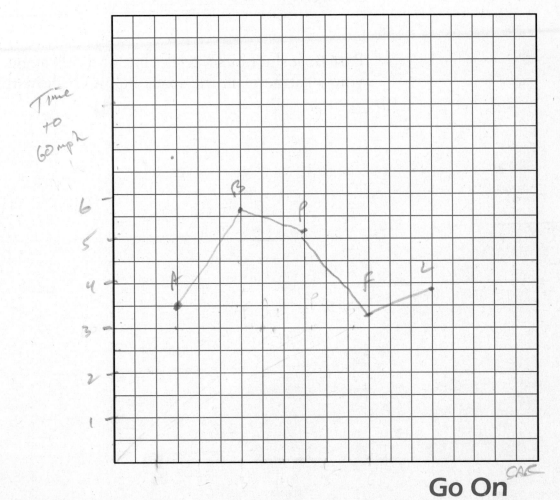

Go On

30. Use your protractor to help you solve this problem. Joshua and Tyler were practicing their trick shots at the pool table, and Tyler hit an angle like the one shown.

Part A: What is the measure of angle *T*?

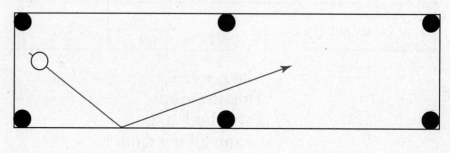

Answer: _____122_____

Part B: Joshua hits a trick shot at a 30° angle. Use your protractor in the space below to show that angle.

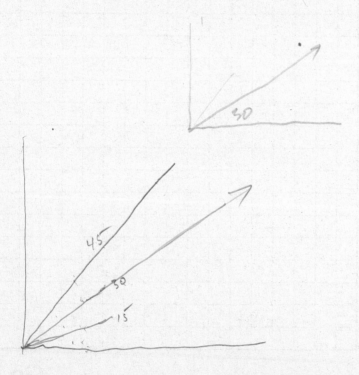

31. Use your ruler to measure the following.

Part A: How many centimeters wide is this "killer bee"?

Answer: _____3_____

Part B: A second bee was found to be twice as wide. Use the space below to draw a line to represent the second bee's width.

Go On

32. With a New Year's Eve party scheduled, the party committee purchased the following amounts of confetti.

Color	Amount of Confetti Purchased
Black	❄❄❄❄❄
White	❄❄❄❄❄❄
Yellow	❄❄❄
Gold	❄❄❄❄❄❄ ❄❄
Silver	❄❄❄❄❄
Red	❄❄❄❄❄❄ ❄❄❄❄❄❄
Blue	❄❄❄❄❄❄

Key
❄ = 20 tons of confetti

Part A: How many tons of blue confetti did the committee purchase?

Answer: ___110___

Part B: The committee then wanted another color. They chose bright pink, and they planned to order it from the company from which they had purchased the first order. How much pink confetti did they order if the amount was greater than the amount of white confetti but less than the amount of gold confetti?

Answer: ___140___

Explain your answer.

_____

_____

_____

_____

_____

_____

_____

_____

_____

_____

Go On

33. Part A: Using your protractor, measure the following angle.

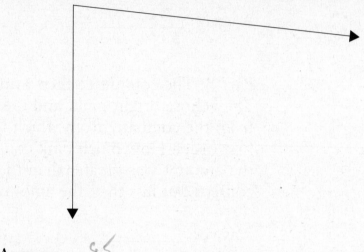

Answer: _____85_____

Part B: What type of angle is it?

Answer: ____Acute____

34. Use your ruler to find the perimeter of the following figure to the nearest centimeter. Label the length of each side.

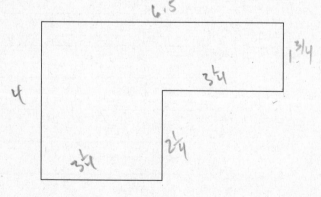

Answer: _____21_____

Stop

ANSWER SHEET:
PRACTICE TEST 2

Book 1

1. Ⓐ Ⓑ Ⓒ Ⓓ
2. Ⓐ Ⓑ Ⓒ Ⓓ
3. Ⓐ Ⓑ Ⓒ Ⓓ
4. Ⓐ Ⓑ Ⓒ Ⓓ
5. Ⓐ Ⓑ Ⓒ Ⓓ
6. Ⓐ Ⓑ Ⓒ Ⓓ
7. Ⓐ Ⓑ Ⓒ Ⓓ
8. Ⓐ Ⓑ Ⓒ Ⓓ
9. Ⓐ Ⓑ Ⓒ Ⓓ

10. Ⓐ Ⓑ Ⓒ Ⓓ
11. Ⓐ Ⓑ Ⓒ Ⓓ
12. Ⓐ Ⓑ Ⓒ Ⓓ
13. Ⓐ Ⓑ Ⓒ Ⓓ
14. Ⓐ Ⓑ Ⓒ Ⓓ
15. Ⓐ Ⓑ Ⓒ Ⓓ
16. Ⓐ Ⓑ Ⓒ Ⓓ
17. Ⓐ Ⓑ Ⓒ Ⓓ
18. Ⓐ Ⓑ Ⓒ Ⓓ

19. Ⓐ Ⓑ Ⓒ Ⓓ
20. Ⓐ Ⓑ Ⓒ Ⓓ
21. Ⓐ Ⓑ Ⓒ Ⓓ
22. Ⓐ Ⓑ Ⓒ Ⓓ
23. Ⓐ Ⓑ Ⓒ Ⓓ
24. Ⓐ Ⓑ Ⓒ Ⓓ
25. Ⓐ Ⓑ Ⓒ Ⓓ
26. Ⓐ Ⓑ Ⓒ Ⓓ

Book 2

Use the answer spaces in the test booklet to provide your answers and show your work.

PRACTICE TEST 2

Book 1: 45 minutes

Directions: Choose the best answer and fill in the corresponding bubble on the answer sheet.

1. Joshua purchased 3 baskets of "sneakers in a basket." How many pairs of sneakers did he have?

$$1 \text{ basket} = 7 \text{ sneakers}$$

   A. 3

   B. 21

   C. 10

   D. 8

2. Use your protractor to measure the angle Jessica drew. What is the measure of angle $\angle C$?

   A. 40°

   B. 70°

   C. 10°

   D. 180°

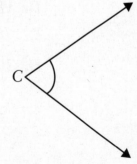

Go On

3. Tyler is one of the best eighth-grade runners. In his last race he ran a 100-yard dash in 12.456 seconds. What is his time rounded to the nearest hundredth of a second?

   A. 12.40 seconds

   B. 12.45 seconds

   C. 12.46 seconds

   D. 12.456 seconds

4. Which of the following is true?

   A. $\frac{1}{3} > 0.2$

   B. $\frac{1}{3} > 0.5$

   C. $\frac{1}{10} > 0.2$

   D. $\frac{1}{10} > 0.5$

5. Jessica purchases an area rug to put in her new bedroom. The rug is 4 feet wide and 6 feet long. What is the perimeter of the area rug?

   A. 9 feet

   B. 18 feet

   C. 24 feet

   D. 20 feet

6. Gabrielle's company has shown the following growth in employees between 2003 and 2008. In what year did Gabrielle's company have the lowest number of employees?

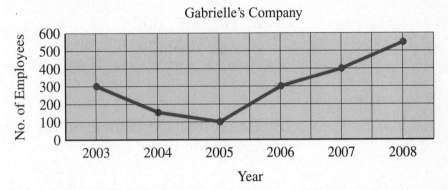

Gabrielle's Company

A. 2008

B. 2006

C. 2005

D. 2004

7. In Rafael's closet 25% of the clothes are shirts. What fraction are shirts?

A. $\dfrac{25}{10}$

B. $\dfrac{25}{1}$

C. $\dfrac{25}{100}$

D. $\dfrac{1}{25}$

Go On

8. Lisa gives the students in her class a problem to find out which of the following sets of angles represents the angles of a triangle. Which is the correct answer?

A. 40°, 40°, 40°

B. 60°, 60°, 60°

C. 90°, 90°, 90°

D. 180°, 180°, 180°

9. Which fraction belongs in the box to make the expression true?

$$\frac{1}{2} < \frac{7}{10} < \square$$

A. $\frac{1}{3}$

B. $\frac{1}{4}$

C. $\frac{4}{5}$

D. $\frac{3}{5}$

10. Which pattern follows the rule given below?

Multiply by 4 and add 2.

A. 10, 42, 170, 682, ...

B. 4, 8, 12, 16, 20, ...

C. 6, 17, 56, 89, 77, ...

D. 2, 4, 6, 8, ...

11. In the latest engineering feat, Tyler's engineering company has developed a new type of roller coaster that uses the following items to build both the track and the carts.

Items	Carts	Track
Lumber (wood)	2,000	40,000
Metal nuts and bolts	4,800	100,000
Pillars	0	4,445

How many metal nuts and bolts are need for the entire project?

A. 2,000

B. 104,800

C. 42,000

D. 4,445

12. Joshua measures the angles of the basement of his house. He notices that the house is built as triangles with two angles measuring 10°. What is the measurement of the third angle?

A. 20°

B. 60°

C. 90°

D. 160°

Go On

13. Which two figures have been divided into equal parts?

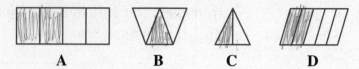

A   B   C   D

BAD PICTURE

  A. figures A and B

  B. figures A and C

  C. figures B and D

  D. figures A and D

14. Triangle *ABC* is congruent to triangle *XYZ*.

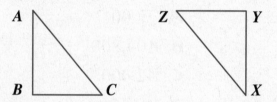

Which parts of the triangles are corresponding?

  A. $\overline{AC}$ and $\overline{ZX}$

  B. $\overline{BC}$ and $\overline{YZ}$

  C. $\overline{BC}$ and $\overline{YX}$

  D. $\overline{AB}$ and $\overline{YZ}$

15. What is the next figure in the pattern?

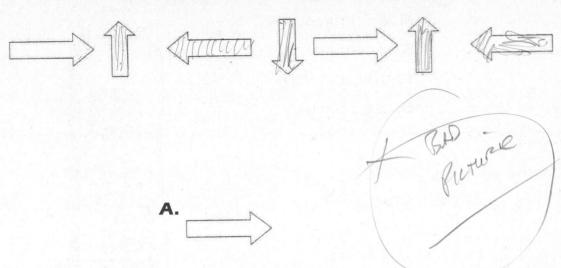

A.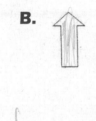

B.

C.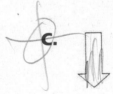

D.

16. Use your ruler to measure the remote control.

   A. 8 centimeters

   B. 4 centimeters

   C. 10 centimeters

   D. 2 centimeters

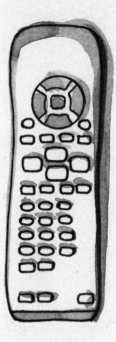

17. U2 gave a concert in Central Park in New York City, and 750,000 people showed up. How many 10 thousands are equal to 750,000?

   A. 750

   B. 7,500

   C. 75,000

   D. 750,000

18. Jessica's scores for the past quarter were as follows: Math, 90; English, 95; Social Studies, 97; Science, 98. What is the mean of Jessica's scores?

   A. 85

   B. 85

   C. 70

   D. 95

19. Tyler's weight before each of his races was as follows.

TYLER'S WEIGHT CHART

Day	Weight (in pounds)
1	110
2	115
3	116
4	112

Which graph represents his weight in the chart?

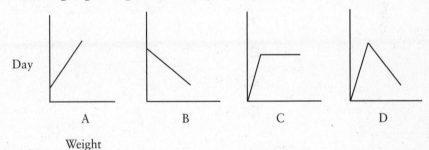

A. graph A

B. graph B

C. graph C

D. graph D

20. Lisa wants to build a fence around a rose garden in order to keep the animals out. The length of the garden is 20 yards, and the width is 30 yards. What is the perimeter of Lisa's garden?

A. 41 yards

B. 50 yards

C. 100 yards

D. 112 yards

Go On

21. The ratio of McIntosh computers to PC computers is 1:3 at the local computer store. Which fraction shows the ratio 1:3?

A. $\frac{1}{3}$

B. $\frac{3}{2}$

C. $\frac{2}{3}$

D. $\frac{1}{2}$

22. What is the least common multiple (LCM) of 7 and 8?

A. 15

B. 56

C. 49

D. 64

23. Your teacher has asked you to place three fractions in the correct order. Which of the following represents the correct order for the fractions?

A. $\frac{2}{3} < \frac{3}{5} < \frac{3}{4}$

B. $\frac{3}{4} < \frac{3}{5} < \frac{2}{3}$

C. $\frac{3}{5} < \frac{2}{3} < \frac{3}{4}$

D. $\frac{3}{5} < \frac{3}{4} < \frac{2}{3}$

24. Simplify the expression below.

$$3 + 8 \times (4 + 2) \div 2 - 1$$

A. 32

B. 27

C. 23

D. 26

25. How many lines of symmetry does an equilateral triangle have?

A. 1

B. 2

C. 3

D. 4

26. Use your protractor to measure the following angle. What is its measurement?

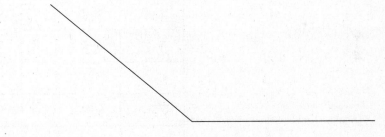

A. 90°

B. 120°

C. 180°

D. 60°

Stop

Book 2: 50 minutes

Directions: Fill in the answers below and show your work.

27. Use the information given to answer the questions below. Joshua is a sophomore in high school and is running track. His coach has asked him to keep track of his daily workout in order to "fine-tune" his running times. Here are Joshua's recorded times for 10 days.

Day	Distance (in miles)	Time (in minutes)
1	4.7	5.12
2	5.8	5.96
3	6.4	6.12
4	5.3	5.12
5	4.5	5.35
6	7.2	6.56
7	5.8	5.16
8	6.2	5.75
9	6.3	5.60
10	6.8	5.40

Part A: Graph the information making sure that you

Use a line graph.
Title the graph.
Provide a scale for the graph.
Label the axes.

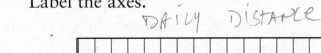

DAILY DISTANCE

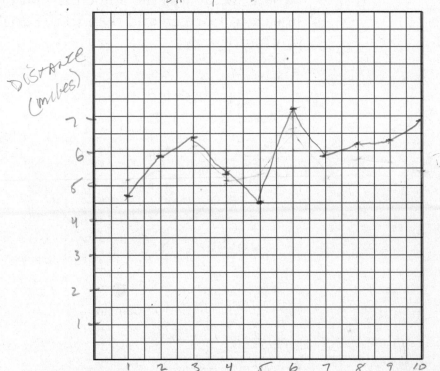

DISTANCE (miles)

DAY

Part B: What was Joshua's mean for the number of miles he ran for the 10 days? During which part of the 10 days did Joshua run the most, the first five days or the second five days? And by how much?

$\overline{m} = 5.9$

2ND 5 BY 5.6 miles

Go On

28. Jessica is the star of her soccer team. She is a team player and passes the ball with ease to her teammates. She is able to score as well. The diagram below shows her passing to Carmen, her teammate. Carmen then scores a goal by shooting the ball into the corner.

Part A: What was the angle the ball traveled from Jessica to Carmen and then to the goal? Use your protractor.

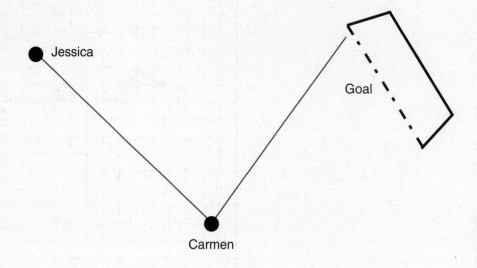

Answer: _____80_____

Part B: What type of angle is the angle formed by the ball's trajectory? Explain.

_____Acute ∠ 90_____

_____

_____

_____

_____

29. In the figure below the shaded triangles are all congruent.

Part A: Find the perimeter if you know the following measurements.

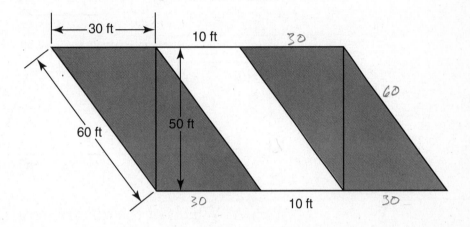

Part B: Explain how you found the perimeter.

_260_

30. Some of the quadrilaterals below have parallel sides.

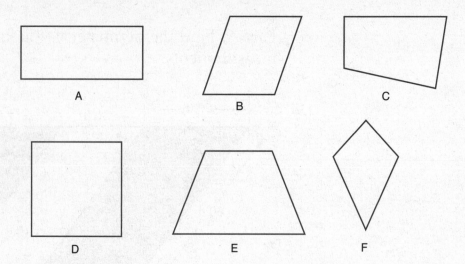

Part A: Which of the quadrilaterals have two pairs of parallel sides?

Answer: _A, B, D_

Part B: What is the name for quadrilaterals with two pairs of parallel sides?

Answer: _PARALLELOGRAM_

Part C: What is the name for quadrilaterals with two pairs of right angles?

Answer: _RECTANGLE_

31. Part A: On the grids below, shade the correct number of squares to represent the decimal 2.21.

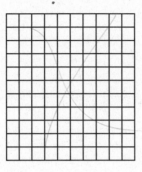 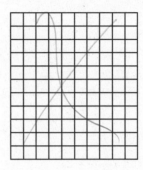 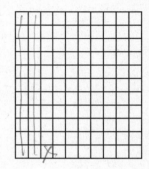

Part B: Round off this decimal to the nearest tenth.

Answer: _2.2_

32. Part A: Write the following fractions in order from the least to the greatest.

$$\frac{1}{10} \qquad \frac{1}{2} \qquad \frac{1}{4} \qquad \frac{1}{3}$$

Answer: _____

Part B: Explain how you ordered the fractions.

_____

_____

_____

_____

_____

Go On

33. Part A: Use your protractor to draw an angle that is 70°.

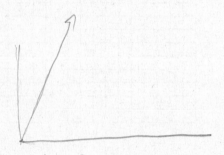

Part B: What type of angle is it? Explain.

acute

34. Alexandra has worked extremely hard to obtain her degree. For her last eight exams in statistics her scores were 96, 90, 97, 96, 99, 98, 98, 89, 98.

Part A: What is the
Mean? _____95.67_____
Mode? _____98_____
Median? _____97_____
Range? _____10_____

9 extras

Part B: Plot a line graph of her exam scores.

89, 90, 96, 96, 97, 98, 98, 98, 99

95.67

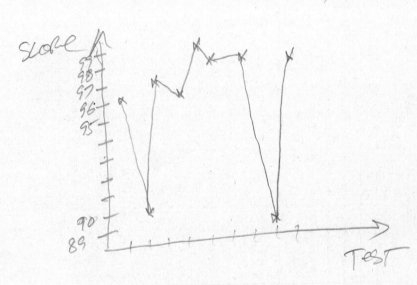

Stop

ANSWER KEY

CHAPTER 1: THE NEW YORK STATE GRADE 5 MATH TEST

ANSWER TO PRACTICE QUESTION 1

Students need to understand that taking the time to read a question carefully and being able to interpret what the question is asking are necessary for success on this test. This question basically asks how to divide 150 (150 clams) by 5 (5 friends) equally. The correct answer is *a* (30).

Some students may answer either *b* (50) or *c* (145), and here are a few reasons that may explain their thinking. Students who chose *c* may think that Sponge Bob has 150 clams and would give just 1 clam to each of his friends, thus satisfying the "*equally*" part of the problem. The other answer students may choose is *b* because they made an arithmetic error—thinking to themselves that $150 \div 5 = 50$.

ANSWERS TO PRACTICE QUESTION 2

Miles on day 1	$15\dfrac{3}{10}$
Miles on day 2	$14\dfrac{1}{10}$
Miles on day 3	$16\dfrac{3}{10}$
Total miles at the end of 3 days	$45\dfrac{7}{10}$ miles
	$45\dfrac{7}{10} - 13\dfrac{4}{10} = 32\dfrac{3}{10}$

In this problem, the student has to realize that in order to calculate the total number of miles Lance Armstrong raced in the 3 days, he or she will have to add all mixed numbers. The first thing students need to ask is, "Do all the mixed numbers have the same denominators?" If so, then they must add all the numerators and keep the denominator 10 the same. When students add the mixed numbers, $45\dfrac{7}{10}$ is the answer.

Therefore, the answer for Part A is $45\dfrac{7}{10}$ miles.

Students need to remember Part B in order to receive complete credit. The question is as follows: One of Armstrong's teammates rode $13\dfrac{4}{10}$ *fewer* miles in the same 3 days. This question asks students to subtract. In other words, $45\dfrac{7}{10} - 13\dfrac{4}{10} = 32\dfrac{3}{10}$. This response is the number of miles that Mr. Armstrong's teammate rode in the same 3 days. For Book 2 responses, if a student makes a calculation error in Part A of the question, the grader is instructed to not penalize the student in their response for Part B.

CHAPTER 2: GEOMETRY

ANSWERS TO PRACTICE QUESTIONS 1

1. A *ray* is a line that begins at a point and never ends. A *line segment* has a beginning and an end. A *line* has neither a beginning nor an end.

   a. ray: $\overrightarrow{JM}$

   b. line segment: $\overline{TM}$

   c. line: $\overleftrightarrow{NM}$

2. Examples of line segments: sides of triangles around the sun, sides of the trucks, rows of crops. Example of rays: the sunrays. Examples of parallel lines: the cut blade of the harvester, sides of the trucks that form a rectangle.

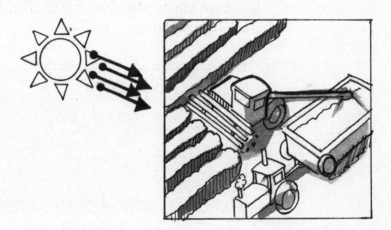

3. There are 10 line segments in the figure: *AD, DC, CB, BA, AE, DE, CE, BE, AC,* and *BD.* Here is the explanation for this problem. Let us verify our answers by looking at drawings piece by piece.

We will now build the figure step by step and count how many line segments we have as we go along.

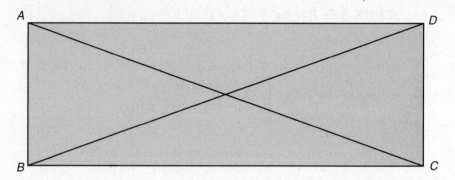

The above figure has 4 line segments, $\overline{AB}$, $\overline{BC}$, $\overline{CD}$, and $\overline{DA}$. Now let us place the other lines, one at a time, over the rectangle. Something we should remember is that we are talking about the same line, $\overline{AB} = \overline{BA}$. It is called the identity property. You do not have to know this for the exam, but it is information for you to have.

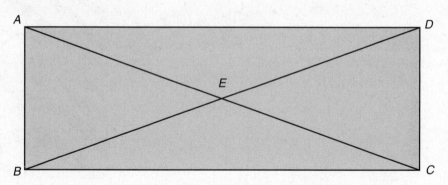

Drawing $\overline{AC}$ and drawing $\overline{BD}$ (in addition to the outside lines) makes it 5 and 6 lines, respectively. Are we done yet? NO, because we also have to count the line segments made by the intersecting line at point E.

Therefore, we have $\overline{AE}$, $\overline{BE}$, $\overline{CE}$, and $\overline{DE}$, which adds 4 more line segments to make 10!

ANSWERS TO PRACTICE QUESTIONS 2

1. The first angle is an acute angle.

2. The middle angle is a right angle.

3. The last angle is an obtuse angle.

Note: sometimes you have to turn your exam paper in order to see which angle you are being asked to name.

ANSWERS TO PRACTICE QUESTIONS 3

1. Remember that the sum of the angles of a triangle equals 180° and that the sum of the angles of a parallelogram equals 360°. Therefore, 180° + 360° = 540°.

2. Let's see:
a. Pentagon—homeplate.
b. Squares—first, second, and third bases and the location of all the bases and homeplate.
c. Circle—pitcher's mound.

3. A regular polygon is a polygon with all sides equal in length.

ANSWERS TO PRACTICE QUESTIONS 4

1. The perimeter can be found by walking around A-Rod's garage and measuring the distance.

$$120 \text{ ft} + 60 \text{ ft} + 120 \text{ ft} + 60 \text{ ft} = 360 \text{ ft}$$

$$(2 \times 120 \text{ ft}) + (2 \times 60 \text{ ft}) = 360 \text{ ft}$$

The correct answer is *c*.

2. The only way to assure yourself as to which one is congruent to rhombus *RAME* is to measure each one. The correct answer is *c*.

3. For problems like this you should use the protractor given to you. But you can eliminate some of the choices by virtue of what you know about triangles. Choice *a* is not possible because the sum of all the angles of a triangle equals 180° and that choice is 200°; *b* cannot be a valid choice because the sum is 115°; and choice *c* is 90°. Therefore, the only choice that is correct is *d*.

4. You will have to use your protractor in the following way in order to get the correct measurement. You will see that the protractor has two sets of numbers, but you know that the angle here is less than 90°; therefore the correct answer is *c*.

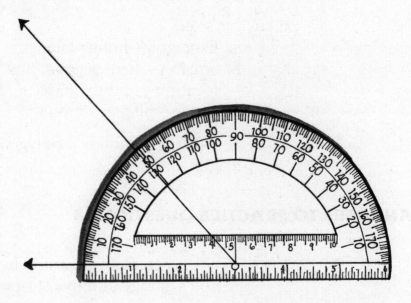

5. Here again you will need to use your protractor. (The rays have been made a little thicker so that you can see the protractor's numbers.) You will see that the angle is close to 145°, which is answer *d*. Notice that sometimes angle measurements of are not exact.

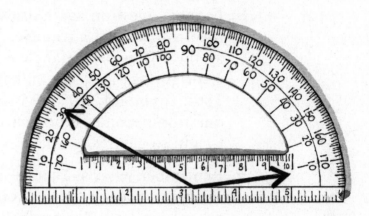

6. Once again you will need your protractor for this question. The correct answer is *d*.

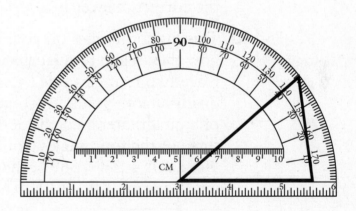

7. For this problem you have to remember the types of angles and their classifications. Acute, less than 90°; right, equal to 90°; obtuse, more than 90° and less than 180°, straight, equal to 180°. Therefore, the correct answer is *c*.

8. For this question again draw on your knowledge of the properties of a triangle. You know that the sum of all the angles of a triangle equals 180°. In this case it is a right triangle, so you have 90° + 15° + $\angle C° =$ 180°. If you add 90° + 15°, you have 105°, and the question becomes how many more degrees you need to add to get 180°. The answer is 75°, answer *b*.

9. Remember what similar means—ratio. Therefore, you are being asked to give the ratio between a small triangle and a big triangle. Start with the smallest side of the smallest triangle, in this case 2, and then find the corresponding side (the side that "goes" with 2), in this case 10. So, you have the ratio $\frac{2}{10}$ or $\frac{1}{5}$, and the correct answer is *a*.

10. This question relies on your knowledge of triangles and how they relate to quadrilaterals. Because you can place two triangles side by side and form a quadrilateral, you know that the sum of all the angles of a quadrilateral equals 360°, and because you have three of the four angles, you have the following: 90° + 90° + 50° + $\angle B$ = 360°. Thus, 230° + $\angle B$ = 360°, which means $\angle B$ = 130°. The only possible answer is *d*.

CHAPTER 3: NUMBER SENSE AND OPERATIONS

ANSWERS TO PRACTICE QUESTIONS 1

1. a. 8 is in the hundreds place.
b. 2 is in the ones place.
c. 6 is in the ten-thousands place.

2. a. $(4 \times 10{,}000) + (6 \times 1{,}000) + (8 \times 100) + (3 \times 10) + (4 \times 1)$

b. $(6 \times 100) + (7 \times 10) + (2 \times 1)$

c. $(7 \times 100{,}000) + (6 \times 10{,}000) + (8 \times 1{,}000) + (0 \times 100) + (4 \times 10) + (5 \times 1)$

ANSWERS TO PRACTICE QUESTIONS 2

1. Digits greater than 4 are 5, 6, 7, and 8.

2. The numbers 0, 1, 2, 3, 4, 5, and 6 are less than 7.

3. a. $6 > 4$

b. $3 < 4$

c. $7 < 8$

d. $8 > 7$

ANSWERS TO PRACTICE QUESTIONS 3

1. If you convert the fractions to decimals you will see that $\frac{1}{6} = 0.167$, $\frac{2}{5} = 0.40$, $\frac{1}{2} = 0.50$, $\frac{4}{7} = 0.57$

2.

3. The four consecutive common multiples of the denominators of $\frac{3}{4}$ and $\frac{2}{3}$ are 12, 24, 36, and 48. The least common multiple is 12.

4. Compare the following fractions

a. $\frac{1}{8} < \frac{1}{2}$

b. $\frac{2}{3} = \frac{6}{9}$

c. $\frac{1}{3} < \frac{4}{5}$

ANSWERS TO PRACTICE QUESTIONS 4

Hundreds	Tens	Ones	Tenths 0.1	Hundredths 0.01	Thousandths 0.001	Ten-thousandths 0.0001	Hundred-thousandths 0.00001	One-millionths 0.000001
1	4	5 .	8	9	3			
		2 .	0	2				
		.	0	7	6			

ANSWERS TO PRACTICE QUESTIONS 5

1. Remember the fractions you were told to memorize? Some of that information can be used here. Since $\frac{1}{4} = 0.25$, and $\frac{1}{3} = 0.33$, choices *c* and *d* are not valid choices. With choice *a*, if you convert $\frac{2}{3}$ to a decimal, that is, 0.67, then you will have that $0.67 > 0.67$, which is false. So the only correct answer is *b*.

2. You will have to remember where the tenths place is and what number is there. In the number 145.962, the number 9 is located in the tenths place. In order to round off, look at the number to the right of this number, and if it is greater than 5, add 1 to the number in the tenths place. In the given number, 6 is the number to the right of the tenths place, and $6 > 5$. So you add 1 to 9, which equals 10. You put 0 in the tenths place and carry the 1. $145 + 1 = 146$. The correct answer is *c*.

3. Remember that 30% means 30 out of 100, answer *b*.

4. For this question ask yourself what 10 thousand looks like. It looks like 10,000. How many zeros do you see? Four zeros—so if you put four zeros after answer *a* you will have 950,000.

5. In this figure there are 3 shaded squares out of 7 total squares. The answer is *b*.

6. A prime number is a number that can be divided only by 1 and by the number itself or a number that has only 1 and itself as factors. Factors of 16 are 1, 2, 4, 8, 16; factors of 2 are 1, 2; factors of 8 are 1, 2, 4, 8; factors of 9 are 1, 3, 9; factors of 7 are 1, 7; factors of 6 are 1, 2, 3, 6; factors of 3 are 1, 3; so only 2, 3, and 7 are prime numbers out of 7 numbers. Therefore, the correct answer is *b*.

7. Since you have common denominators, just add the numerators. The correct answer is *c*.

CHAPTER 4: PROBLEM SOLVING

ANSWERS TO PRACTICE QUESTION 1

Think about the strategies.

1. Do you understand the problem?
   a. What is the relevant (important) information:
      i. You are saving money each month (I hope this is the case because a little adds up) for 4 months.
      ii. Each month you save an amount of money.
      iii. The state exam may ask you to
         (1) Title your graph.
         (2) Label both axes.
         (3) Graph all the data.
         (4) Provide a scale for the graph.
         Therefore, it will be important for you to cross out each one of these as you begin to attack the problem.

b. Extraneous (more than necessary) information:
   i. Money you save from doing chores and from birthday and Christmas gifts. (At least for the question that was asked, it would be important information for you to know to send thank-you cards to people who sent you money.)
   ii. Even the specific months are not important in solving the question.

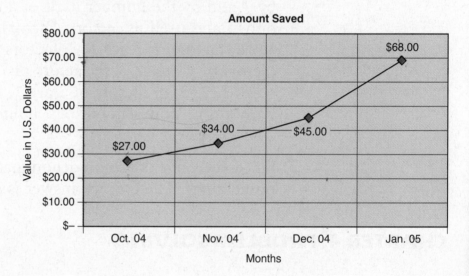

Your histogram may look like this one.

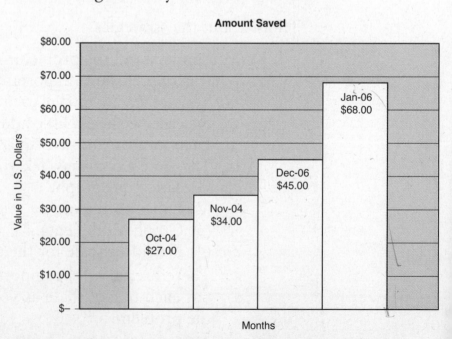

ANSWERS TO PRACTICE QUESTIONS 2

1. There are a few things you have to remember. Read through the problem and see if it makes sense to you. What do you think of when you read the word "total"? You might think of a sum or of adding numbers to obtain a bigger number than the one you started with (at least for our purposes; later in your mathematical career you will experience negative numbers, but for now "total" means more than what you started with). When answering this question, do not just put the answer down, show your work.

Part A: Pedro ran for 4 days, each day clocking a different distance. We have to add these distances together:

$$5\frac{3}{4} \text{ km} + 4\frac{1}{4} \text{ km} + 6\frac{2}{4} \text{ km} + 4\frac{3}{4} \text{ km}$$

If you add the whole numbers first, you will get the following.

$$5 + 4 + 6 + 4 = 19$$

Now add the fractions:

$$\frac{3}{4}+\frac{1}{4}+\frac{2}{4}+\frac{3}{4}=\frac{9}{4} \quad \text{or} \quad = 2\frac{1}{4}$$

Adding the whole numbers and the fractions,

$$19+2\frac{1}{4}=21\frac{1}{4}$$

The answer is $21\frac{1}{4}$ kilometers.

Part B: Since we know that Pedro ran $21\frac{1}{4}$ kilometers, all we need to do is add that amount to Janet's additional distance:

$$4\frac{2}{4} \text{ km} + 21\frac{1}{4} \text{ km} = 25\frac{3}{4} \text{ km}.$$

The answer is $25\frac{3}{4}$ kilometers.

2. For the graph you will need to start by labeling both axes (x, y). The x-axis will be for the dates of the stock market, and the y-axis for the value of the stocks in dollar amounts. It should look something like this.

Dates of Stock Market

Then you will have to decide what scale you need to use to show the values of the stocks. One thing you should consider is that if the numbers are large, you cannot have a scale that increases by small amounts.

The opposite is also true; if you have small numbers, you do not want to use too large a scale. Notice that in this problem the smallest number is 52.25 and the largest is 65.87. Therefore, if you choose to begin at 50.00 and end at 68.00, it will give the reader of the graph a good idea of what the information is. Notice also that you need to display everything you are asked for. Here we have provided a title for the graph, labeled both axes, graphed all the data, and provided a scale for the graph.

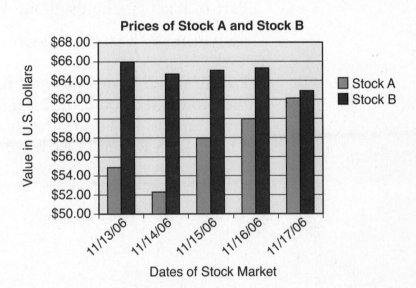

3. Part A:

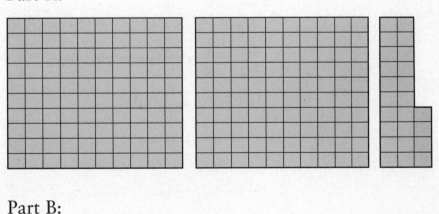

Part B:

2.16   2.34   2.5

4. Part A: Because the question asks you to estimate, an exact answer is not required. Look at your data ($\approx$ means "about"):

> 4.7 jars $\approx$ 5 jars
> 3.75 jars $\approx$ 4 jars
> 5.5 jars $\approx$ 6 jars
> 3.25 jars $\approx$ 3 jars

So there are about 18 jars.

Part B: Each jars holds about 45 worms, so

45 worms $\times$ 18 jars = 810 worms

Part B: If you have 5 jars and Bruce Wayne has 6 jars, then combined you have 5 + 6 jars, or 11 jars. Because every jar has 45 worms, the total number of worms is 11 jars $\times$ 45 worms in each jar, or 495 worms.

5.
$$2 \quad 2\frac{1}{8} \quad 2\frac{1}{4} \quad 2\frac{3}{8} \quad 2\frac{1}{2} \quad 3$$

The explanation could be something like this: $2\frac{1}{2}$ has to go between numbers 2 and 3, $2\frac{1}{4}$ is between 2 and $2\frac{1}{2}$, $2\frac{1}{8}$ is halfway between $2\frac{1}{4}$ and 2, and $2\frac{3}{8}$ is between $2\frac{1}{4}$ and $2\frac{1}{2}$.

CHAPTER 5: MEASUREMENTS

ANSWERS TO PRACTICE QUESTIONS 1

1. Make sure you use the stairs.
   a. 100 cm   *10*
   b. 10 dm   *1*
   c. 1,000 mm   *100*

2. Metric units
   a. m
   b. dm
   c. dm
   d. km

3. The perimeter is the total distance around a figure. You have a rectangle, and by definition, you know that the opposite sides of a rectangle are parallel and equal; therefore, the side opposite 15 m will be 15 m, and the side opposite 60 m will also be 60 m. If you are given 35 m + 5 m for this opposite side, and you know it must be 60 m in total, then the remainder is 20 m. The other important information is that the second figure in the diagram is a square, so all of its sides must be equal, 5 m. The perimeter of the entire figure (rectangle plus square), therefore, is 15 m + 15 m + 60 m + 60 m + 15 m (for the three sides of the square not included in the rectangle measurement), which adds up to 165 m.   *160 double count. open side of square*

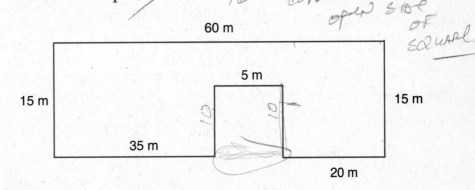

4. If you count from 11:25 a.m. forward until 1:25 p.m., you will see that this is 2 hr, and from 1:25 until 1:26 is 1 min. So the time elapsed was 2 hr and 1 min.

5. Unless you live next door to the mall, it is not a appropriate unit to use. It would make more sense if it were 120 km because kilometers are used when dealing with distances that you travel. Another way of thinking about this is to realize that 260 m = 284 yd, which is about two and one-half football fields.

6. Refer to the time chart within the chapter to do this problem.

$$110 \text{ yr} \times \frac{365 \text{ days}}{1 \text{ yr}} = 110 \times 365 \text{ days}$$
$$= 40{,}150 \text{ days}$$

What if I asked how many hours this would be? Can you figure it out?

$$40{,}150 \text{ days} \times \frac{24 \text{ hr}}{1 \text{ day}} = 963{,}600 \text{ hours old. Now that}$$

is old.

7. Remember to use the stairs. How many steps are there from centimeters to meters? There are 3 steps, which means using 10^3; and since you are going down, you will be multiplying—meaning that you will move the decimal to the right. In other words, $0.25 \times 10^3 = 250$ cm.

$10^2 \quad 25$

$100 \, cm = 1 \, m$

CHAPTER 6: PATTERNS AND FUNCTIONS

ANSWERS TO PRACTICE QUESTIONS 1

There are two ways of doing all these problems. We will show you both. Try them both and see which one you feel most comfortable with.

1.

$$n \xrightarrow{+\ 16} 51$$
$$\xleftarrow{-16} \quad n = 35$$

$$16 + n = \boxed{51}$$
$$\underline{-16 \quad\ -16}$$
$$n = 35$$

2. $w = 5$

3. $w = 26$

4. $e = 1{,}010$

5. $r = 0$

ANSWERS TO PRACTICE QUESTIONS 2

1. $\$85.00 - \$35.95 = b$; so $b = \$49.05$

2. Let g = my great-grandmother's age and m = my age.

$g = 2m + 5$

3. Let a = input and b = output.

$b = 2a + 4$

4.

5. 28, 20

6. You can do this in either of the ways you were shown.

a. $n + 6 = 14$
$$-6 = -6$$
$$n = 14 - 6$$
$$n = 8$$

b. $3x + 5 = 14$

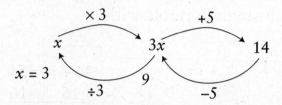

Explanation: You start with the variable x. When you multiply it by 3, you get $3x$; then add 5, which results in 14. In order to find the answer, you have to work backward from 14 toward the variable. Along the way ask yourself, What did I do to get to the point where I am now? You begin with 14 and do the opposite; because the last thing done was adding 5, you have to subtract 5 from 14, which equals 9. Next ask yourself what was done on the way to this point, and the answer is that you multiplied by 3. Therefore, you need to divide 9 by 3, and the result is that $x = 3$.

c. We have the following:
$$44 - r = 12$$
$$+ r = \quad + r$$
$$\overline{}$$
$$44 \quad = 12 + r$$
$$-12 \quad -12$$
$$\overline{}$$
$$32 \quad = r$$
$$r = 32$$

7. In order to graph this equation the value for *y* needs to be found.

x	*y*	*(x, y)*
0	3(0) + 4 0 + 4 4	(0,4)
1	3(1) + 4 3 + 4 7	(1,7)
2	3(2) + 4 6 + 4 10	(2,10)
3	3(3) + 4 9 + 4 13	(3,13)
4	3(4) + 4 12 + 4 16	(4,16)
5	3(5) + 4 15 + 4 19	(5,19)

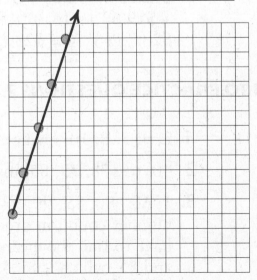

8. The pattern is from a three-sided figure to a four-sided figure to a five-sided figure. The next figure is a seven-sided figure, a heptagon.

9. Let *b* represent Luis' age and let *s* represent my sister's age.
   a. $b + 1$
   b. $s + 5$

CHAPTER 7: DATA AND PROBABILITY

ANSWERS TO PRACTICE QUESTIONS 1

1. $35 + 45 + 55 + 63 = 198$
   $198 \div 4 = 49.5$ points per game

2. $88 + 79 + 85 + 90 + 92 + 95 + 95 = 624$
   $624 \div 7 = 89.14$ words per minute

3. $5 + 6 + 5 + 5 + 4 + 4 + 5 = 34$
   $34 \div 7 = 4.86$ hours per day

4. $90 + 89 + 95 + 96 + 97 + 99 + 95 = 661$
   $661 \div 7 = 94.43$ is your average up to now in your math class.

5. $\$12.25 + \$52.20 + \$15.75 + \$60.25 + \$10.15 + \$10.15 + \$19.68 + \$16.30 + \$19.62 + \$19.63 = \$235.98$
   $\$235.98 \div 10 = \$23.60$ per month

ANSWERS TO PRACTICE QUESTIONS 2

1. 2, 4, 5, 8, 10; median = 5

2. 40, 65, 78, 79, 85; median = 78

3. 1, 60, $\boxed{80, 85}$, 90, 98; because there is an even number count, in order to find the median we will have to find the mean of the two middle numbers, 80 and 85, which is 82.5. Therefore, the median = 82.5.

4. 737, 747, 767, 777, 787; median = 767.

5. 1, 2, $\boxed{5, 7}$, 8, 9; again in this data set the count of numbers is even, so we need to find the mean of the data set in order to find that the median = 6.

ANSWERS TO PRACTICE QUESTIONS 3

1. <u>3</u>, 4, 5, 4, <u>3</u>, <u>3</u>, <u>3</u>, 5, 7: mode = *3*

2. 10, <u>12</u>, <u>12</u>, 5, 5, 8, 1, 6, <u>12</u>; mode = 12

3. 5, 8, 4, <u>0</u>, 1, 10, <u>0</u>; mode = 0

4. 80, 40, 50, 90, 50, 80; mode = none

ANSWERS TO PRACTICE QUESTIONS 4

1. 5, 7, 8, 9, 10, 11, 12: range = 7

2. 10, 50, 65, 79, 89, 98 ; range = 88

3. 44, 50, 58, 90, 100; range = 56

4. 950, 1,000, 1,200, 1,250, 1,300, 1,350; range = 400

5. 78,000; 80,000; 82,000; 83,000; 85,000; 89,000; range = 11,000

ANSWERS TO PRACTICE QUESTIONS 5

1.   i. 16, 19, 40, 52, $\boxed{52, 55}$, 60, 60, 60, 78
   a. mode = 60
   b. median = $\dfrac{52 + 55}{2}$ = 53.5
   c. mean = 49.2
   d. range = 62

ii. 25, 27, 30, 32, $\boxed{35}$, 43, 43, 45, 50
   a. mode = 43
   b. median = 35
   c. mean = 36.66
   d. range = 25

iii. 0, 50, 55, $\boxed{60}$, 70, 80, 90
   a. mode = none
   b. median = 60
   c. mean = 57.86
   d. range = 90

iv. 5, 16, 19, 20, 21, 21, 33
   a. mode = 21
   b. median = 20
   c. mean = 19.29
   d. range = 28

2.

Food	Tally
Fish fry	III
Buffalo chicken wings	III
Cheese pizza	III
Spanish rice and beans	IIII
Salad	I
Spanish hot dog	I

3. a. 3 students drank milk and 7 students drank water. In order to find out how many more students drank water, you subtract: 7 − 3 = 4.

   b. In the instructions for the graph, you were told that all the students in Ms. Estrella's class participated. Therefore, you add: 5 + 12 + 7 + 3 = 27.

4.

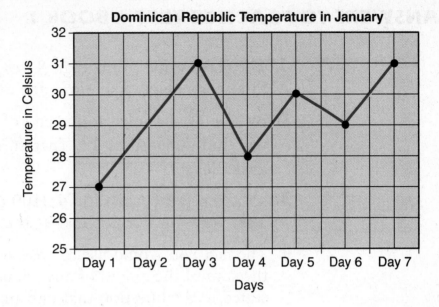

Dominican Republic Temperature in January

5. a. White, 1
Light blue, 5
Dark blue, 4

b. There are 10 triangles in Felix the Cat's bag.

c. $p(\text{light blue}) = \dfrac{5}{10}$

d. $p\ (\text{not light blue})\ \dfrac{5}{10}$

e. That light blue is the most likely to be chosen is because there are five light blue triangles and it is the most common color in Felix the Cat's bag.

ANSWERS TO PRACTICE TEST 1: BOOK 1

1. The place value of the digit 8 is 8,000 since 8 is in the thousands place. Therefore, the correct answer is *b*.

2. By placing the ruler at one end of the guideline, you will find it measures $2\frac{1}{2}$. Therefore, the correct answer is *c*.

3. 45% represents the ratio 45:100 or 45 out of 100. Therefore, the correct answer is *c*.

4. The fact you would use to answer this question is that the sum of the angles of any triangle equals 180°. It is stated in the question that one angle is 30° and the other 20°, and if you add these two angles, you will have 50°. Then the question is what other angle added to 50° will equal 180°, and the answer is 130°. Therefore, the correct answer is *c*.
$$30° + 20° = 50°$$
$$180° - 50° = 130°$$

5. Since a ratio is another way of representing a fraction, where the first number is the numerator and the second number is the denominator, 2:5 represents $\frac{2}{5}$. Therefore, the correct answer is *d*.

6. This question asks you to think about what happens in a car that accelerates (meaning that it picks up speed) and then is constant (meaning that the speed stays the same). Graph A represents a car that speeds up and then decelerates at the same rate or stops. Graph C represents a car that continues to speed up. Graph D represents a car that remains at the same speed. Therefore, the correct answer is *b*.

7. Since each bag has 6 marbles and you have 72 marbles altogether, if you count 6 marbles in one bag and 6 more marbles in a second bag, you are left with 60 marbles. Notice that the pattern is that you are dividing the marbles into bags, and since you have 72 marbles altogether and 6 marbles in each bag, the correct answer is *a*.

8. This question asks you to estimate, and there are several ways of doing this. You can take the $5.00 and subtract $1.20 from it each time until nothing or close to nothing is left:

5.00 – 1.20 = 3.80 (one car)
3.80 – 1.20 = 2.60 (two cars)
2.60 – 1.20 = 1.40 (three cars)
1.40 – 1.20 = 0.20 (four cars)

Or you can think to yourself, "If the cars were cost $1.00 each, I would be able to by 5 cars, but each car costs an extra $0.20, and therefore, I can only buy less than 5 cars. Since I know that if each car was worth $1.20, and since there are 4 quarters in a dollar, I will be able to buy 4 cars. The correct answer is *c*.

9. In this question think about what value the digit 1 occupies—the hundreds place or 100. You are asked how it would change, so answer *a* is incorrect. If we place 7 in the hundreds place, then its value will be 700, and so answer *b* is incorrect. You are asked for the CHANGE in value, so answer *d* is incorrect. If you keep following this logic, you will see that the only correct answer is *c*. You could say that if the value of 7 is 700 and the value of 1 is 100, the change will be 700 – 100 = 600, answer *c*.

10. A hummingbird flapping its wings 120 times in 1 second represents a 120:1 ratio. If we need to know how many flaps, f, occur in 5 seconds, that ratio would be f:5. So we can set it up as follows: $\frac{120}{1} = \frac{f}{5}$. Cross-multiplying, $f = 120 \times 5$ or $f = 600$. The correct answer is c.

11. In order to find how much framing is needed for the picture, you will need to find the perimeter, which is

$$10 + 22 + 10 + 22 = 64$$

Therefore, the correct answer is b.

12. The rule here is to add 250 to the previous amount. So

$250 + 250 = 500$
$500 + 250 = 750$
$750 + 250 = 1,000$

Therefore, the correct answer is b.

13. If you count from 5:15 until 6:15, it is an hour. From 6:15 until 7:15 is another hour. From 7:15 until 7:45 is 30 minutes. Therefore, the correct answer is c.

14. The only answer that is reasonable is 15 meters, answer c.

15. This question asks you to calculate the number of home runs Mr. Mercado hit in 4 days, and you knows that each ball represents 2 home runs. Therefore, because there are 10 balls and each ball represents 2 home runs, the only correct answer is b.

16. Finding the LCM is the same as finding what multiples two numbers have in common and choosing the least (smallest). Therefore, the correct answer is b. Answers c and d are multiples as well, but only answer b is the least multiple.

17. The ratio of the height of the tree to the height of the building is 10:65 or $\frac{10}{65}$. Therefore, the correct answer is *c*.

18. The ratio of number of jumps to the number of seconds is 16:10; therefore, $\frac{16}{10} = \frac{j}{60}$. If you

cross-multiply, $j = \frac{16 \times 60}{10} = 96$.

Therefore, the correct answer is *d*.

19. If you place any of these answers in the open sentence, only answer *a* makes the sentence true.

20. Place the numbers in the data sample in order: 25, 32, 45, 55, 60, 80. The range is $80 - 25 = 55$, so the correct answer is *d*.

21. Think about the numbers 1, 2, 3, 4, 5, 6, 7, 8, 9, 10, 11, and 12. Which of these are multiples of 3? 3, 6, 9, and 12. So there are 4 such numbers out of the total of 12; therefore, the correct answer is *c*.

22. The only correct estimation for the height of an adult is answer *a*.

2 meters = about 6 feet
20 meters = about 65 feet
7 meters = about 23 feet
10 meters = about 33 feet

23. If you do these problems mentally or know about the distributive law of multiplication over addition, you will find that the correct answer is *c*.

a. $4 \times (2 + 5) \overset{?}{=} (4 \times 2) + 5$ b. $4 \times (2 + 5) \overset{?}{=} (4 \times 2) \times 5$
     4×7 $8 + 5$ 4×7 8×5
       28 $\neq$ 13 28 $\neq$ 40
d. $4 \times (2 + 5) \overset{?}{=} (4 + 2) \times (4 + 5)$
     4×7 6×9
       28 $\neq$ 54

24. Each square in the grid represents 0.01; each vertical strip represents 0.1, and there are 8 such strips, which adds up to 0.8. Another way of thinking about this is that if there are 10 vertical strips and only two of them are not colored, you have 8 and each strip represents 0.1. Therefore, the correct answer is *c*.

25. If you count the squares, you will count 50. Or, note that there are 5 squares vertically and 10 squares horizontally, and since you know that area equals length times width,

$$a = l \times w$$
$$= 10 \times 5$$
$$= 50$$

Therefore, the correct answer is *c*.

26. The pattern is three figures that repeat themselves with every other one being shaded, so the next figure will NOT be shaded and will be a diamond. Therefore, the correct answer is *b*.

ANSWERS TO PRACTICE TEST 1: BOOK 2

27. The way to approach this problem is to think of the games each being $7.00, because $7.00 is a good estimate for $6.95. Then multiply by the 7 games you are buying. So the answer is about $49.00.

28. The way to approach this problem is with the knowledge that the sum of the angles in a triangle equals 180°. You have 52° + 22° = 74°; therefore, 180° − 74° = 106°.

29. A line graph could be similar to the following.

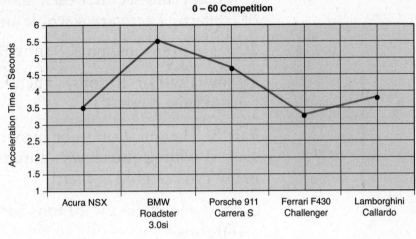

30.

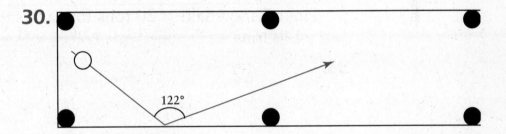

31.

32. Part A: The way to approach this question is to look at the key and see that each snowflake equals 20 tons of confetti. There are 5 whole snowflakes, which is $20 \times 5 = 100$ tons, plus $\frac{1}{2}$ snowflake, which equals 10 tons (half of 20 tons). Therefore, $100 + 10 = 110$ tons of blue confetti.

Part B: The question to be answered is how many tons of white confetti was purchased, as well as how many tons of gold confetti.

Gold: 8 snowflakes × 20 tons for each snowflake = 160 tons

White: 6 snowflakes × 20 tons for each snowflake = 120 tons

Pink: 7 snowflake × 20 tons for each snowflake = 140 tons

33.

Part A: The angle measure is 85°.

Part B: Because this angle measures less than 90°, it is an acute angle.

34. Use your ruler to measure the perimeter. (Remember that the perimeter of a figure is the length around the figure.)

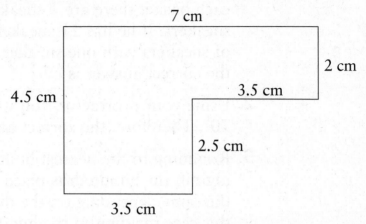

4.5 cm + 7 cm + 2 cm + 3.5 cm + 2.5 cm + 3.5 cm = 23 cm

The perimeter is 23 cm.

ANSWERS TO PRACTICE TEST 2: BOOK 1

1. Since Joshua purchased 3 baskets of sneakers and in each basket there are 7 sneakers, he has $3 \times 7 = 21$ sneakers. If he has 21 sneakers, then he has 10 pairs of sneakers with one sneaker remaining. Therefore, the correct answer is *c*.

2. Using your protractor, you will find the angle to be 70°. Therefore, the correct answer is *b*.

3. Rounding to the nearest hundredth means that the digit in the hundredths place will change or remain the same depending on the digit to the right of it. In this case the digit in the hundredths place is 5. The digit to the right of 5 is 6, and since 6 is *greater* than 5, we add 1 to the digit occupying the hundredths place. Therefore, the correct answer is *c*.

4. The first choice, 0.3 > 0.2, is true. Therefore, the correct answer is *a*.

5. If you draw the rug, it will look like this:

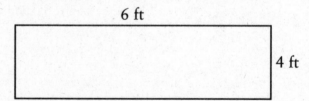

6 ft

4 ft

Since the perimeter is the sum of measurements around the figure, we have that the perimeter of the rug

= 2 (6 ft) + 2 (4 ft)
= 12 ft + 8 ft
= 20 ft

Therefore, the correct answer is *d*.

6. If you look at the graph, you will see that the lowest point shows the lowest number of employees in 2005. Therefore, the correct answer is *c*.

7. Percent means out of 100. Therefore, the correct answer is c, $\dfrac{25}{100}$.

8. For this question you have to remember that the sum of the angles of a triangle equal 180°. So the correct answer will be the three angles that when added together give 180°. Therefore, the only correct answer is b.

9. In one of the chapters in this book, it was suggested that you memorize a few fraction-decimal equivalents in order to make questions like this one easier. Here you have to know the decimal value of each of the fractions: $\dfrac{1}{2} = 0.5$; $\dfrac{1}{4} = 0.25$; $\dfrac{1}{5} = 0.20$; $\dfrac{1}{3} = 0.33$. With these you can calculate the value of each of the given fractions. Because $\dfrac{4}{5} = 0.8$, the only correct answer is c.

10. The only answer that follows the rule is a.

11. For questions involving charts, make sure you are tracking what you are asked to answer. If you want to know the number of nuts and bolts needed for the *entire* project, then it's just a matter of adding 4,800 + 100,000 for both the carts and the track, which equals 100,000 + 4,800 = 104,800. Therefore, the correct answer is b.

12. This problem asks you to remember that the sum of the angles of a triangle equals 180°. You already know that

$$10° + 10° + ? = 180°.$$

Therefore, you are looking for the third angle to be

$$20° + ? = 180° \quad \text{or} \quad ? = 180° - 20°$$
$$= 160°, \text{answer } d.$$

13. Figure A is shaded a half, figure B is shaded a third, figure C is shaded a half, and figure D is shaded three-fifths. Therefore, the only correct answer is b.

14. When figures are congruent—meaning they are equal in measure—all the parts that can be placed on top of each other are called corresponding. Therefore, the correct answer is *b*.

15. The figures repeat themselves as well as they are shaded. The first is not shaded, the next three are, the next one of that is also not shaded. The only correct answer is *b*.

16. Using your ruler, you will see that the remote control measures 8 centimeters, answer *a*.

17. This question asks you to find out which one of the possible answers equals 750,000 when multiplied by 10. In other words,

    $10 \times ? = 750,000$
    $750,000 \div 10 = 75,000$

    Therefore, the correct answer is *c*.

18. The sum of her scores is $90 + 95 + 97 + 98 = 380$

    $380 \div 4 = 95.$

    Therefore, the correct answer is *d*.

19. The only graph that shows an increase and then a decline is graph D, answer *d*.

20. The correct answer is *c*.

    The perimeter is 2(30 yards) + 2(20 yards)
    $$= 60 \text{ yards} + 40 \text{ yards}$$
    $$= 100 \text{ yards}.$$

21. The correct answer is *a*.

    Because the ration of 1 : 3 is given the answer is $\frac{1}{3}$.

22. The LCM represents the first time in the multiplication table that both digits share a common number. The correct answer is *b*.

23. Calculate the decimal equivalent of each fraction by dividing the numerator by the denominator. You will see that the only correct answer is *c*.

24. To find the order of operations, remember

PEMDAS (Please Excuse My Dear Aunt Sally), which stands for parentheses, exponent, multiplication, division, addition, subtraction.

$3 + 8 \times (4 + 2) \div 2 - 1$
$3 + 8 \times (6) \div 2 - 1$
$3 + 48 \div 2 - 1$
$3 + 24 - 1$
$27 - 1 = 26$

Therefore, the correct answer is *d*.

25. Because you are dealing with an equilateral triangle (all angles and all sides are equal), you have three lines of symmetry, *c*.

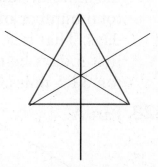

26. The measure of this angle is 120°, answer *b*.

ANSWERS TO PRACTICE TEST 2: BOOK 2

27. Part A:

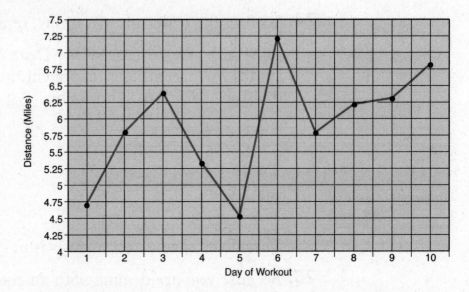

Part B: Joshua's mean number of miles is the sum of the miles divided by the number of days he ran. The total number of miles he ran was 59.2 miles in 10 days. That is 5.9 mean miles run for those days. The first 5 days he ran 26.7 miles; the second 5 days he ran 32.5 miles.

28. Part A:

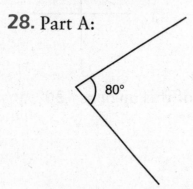

Part B: The trajectory of the ball forms a acute angle. Remember that there are four types of angles: Acute— less the 90°, right—equal to 90°, obtuse—more than 90° but less than 180°, and straight—equal to 180°.

29. In order to answer this question you will need to find the measurements of the entire figure. You are told that all shaded triangles are congruent. Therefore, you need to know that the corresponding parts of the congruent triangles are congruent themselves.

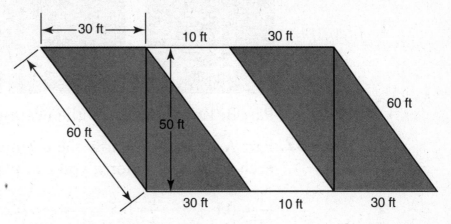

Part A: You have

Top:	30 ft + 10 ft + 30 ft = 70 ft
Bottom:	30 ft + 10 ft + 30 ft = 70 ft
Left side:	60 ft = 60 ft
Right side:	60 ft = 60 ft
Total	= 260 ft

Part B: This can serve as the explanation along with the fact that congruent triangles have congruent parts.

30. Part A: Having two pairs of parallel sides means that two different sets of sides are parallel. The only figures that fit this definition are figures A, B, and D. Figure C does not have any parallel sides, figure E has only one set of parallel sides, and figure F has none.

Part B: Quadrilaterals that have two pairs of parallel sides are called parallelograms.

Part C: Quadrilaterals with two sets of right angles are called rectangles (remember that a square is a rectangle with all sides equal).

31. Part A: Shade two entire "blocks" and then two columns and one square.

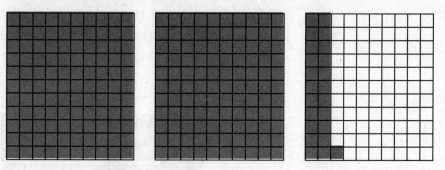

Part B: Rounding off to the nearest tenth gives 2.2.

32. Part A: Once you know the decimal equivalent of each of these fractions, you can place them in order.

$$\frac{1}{10} = 0.1 \qquad \frac{1}{2} = 0.5 \qquad \frac{1}{4} = 0.25 \qquad \frac{1}{3} = 0.33$$

Therefore, the order is

$$\frac{1}{10}, \quad \frac{1}{4}, \quad \frac{1}{3}, \quad \frac{1}{2}$$

Part B: The way to explain how you got your answer is to say that you converted the fractions to decimals to find the equivalent of each one. Just a reminder: As suggested earlier, memorizing some of the basic fractions would make life much easier.

33. Part A:

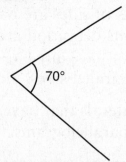

70°

Part B: Since 70° is less than 90°, it is a acute angle.

34. First, you must place the data set in order, least to greatest will do:

89, 90, 96, 96, 97, 98, 98, 98, 97

Mean: 89 + 90 + 96 + 96 + 97 + 98 + 98 + 98 + 97 = 859. Dividing by the number of data in the set (9) gives 95.44.

Mode: 98—it appears three times

Median: 97—it is the middle number

Range: 98 – 89 = 9

Part B:

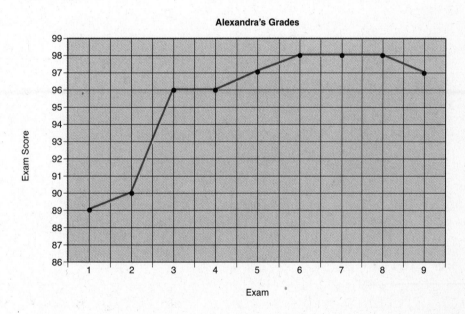

Alexandra's Grades

INDEX

BARRON'S prepares students for the Grade 5 New York State Assessment Tests!

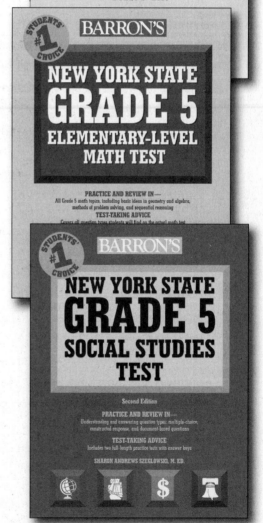

BARRON'S NEW YORK STATE GRADE 5

English Language Arts Test, 2nd Edition

The contents of this revised second edition are based on previous State testing and are consistent with the National Common Core Standards for English Language Arts. This book presents practice exercises and review sections for all of the test's question types. Students will also profit from the book's two full-length practice tests with answers explained plus editing exercises designed to improve their performance on written response questions. Helpful additional features include a glossary of standards-based ELA terms, and a detailed explanation of the test's structure, so students will know exactly what to expect on test day.

ISBN: 978-1-4380-0192-0, $12.99, *Can$14.99*

BARRON'S NEW YORK STATE GRADE 5

Elementary-Level Math Test

This detailed math review guides fifth-grade students through the fundamentals of sequential reasoning and problem solving, and introduces them to number sense and the basic principles of algebra and geometry. The book also features practice-and-review questions with answers plus two full-length practice exams that are similar in format and question types to the actual New York State test. The practice exams come with answers for all questions.

ISBN: 978-0-7641-3945-1, $12.99, *Can$15.99*

BARRON'S NEW YORK STATE GRADE 5

Social Studies Test, 2nd Edition

This updated manual prepares New York State fifth graders for the required statewide social studies test. The author provides a detailed explanation of the test's contents and its different question types, which include multiple-choice questions; constructed response questions, which require brief answers; and document-based questions, each of which requires a short essay. She explains each question type with a series of examples based on fifth-grade-level social studies material, and she offers test-taking advice on how to answer the questions. The manual concludes with two full-length practice tests with answer keys and a vocabulary list of social studies-related words.

ISBN: 978-0-7641-4025-9, $12.99, *Can$15.99*

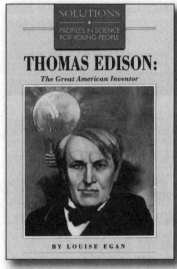
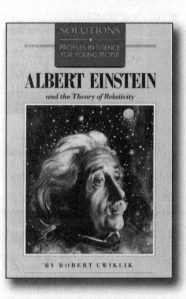